Advances in

Geosciences

Volume 31: Solid Earth Science (SE)

ADVANCES IN GEOSCIENCES

Editor-in-Chief: Kenji Satake *(University of Tokyo, Japan)*

A 5-Volume Set

Volume 1: Solid Earth (SE)
ISBN-10 981-256-985-5

Volume 2: Solar Terrestrial (ST)
ISBN-10 981-256-984-7

Volume 3: Planetary Science (PS)
ISBN-10 981-256-983-9

Volume 4: Hydrological Science (HS)
ISBN-10 981-256-982-0

Volume 5: Oceans and Atmospheres (OA)
ISBN-10 981-256-981-2

A 4-Volume Set

Volume 6: Hydrological Science (HS)
ISBN 978-981-270-985-1

Volume 7: Planetary Science (PS)
ISBN 978-981-270-986-8

Volume 8: Solar Terrestrial (ST)
ISBN 978-981-270-987-5

Volume 9: Solid Earth (SE), Ocean Science (OS) & Atmospheric Science (AS)
ISBN 978-981-270-988-2

A 6-Volume Set

Volume 10: Atmospheric Science (AS)
ISBN 978-981-283-611-3

Volume 11: Hydrological Science (HS)
ISBN 978-981-283-613-7

Volume 12: Ocean Science (OS)
ISBN 978-981-283-615-1

Volume 13: Solid Earth (SE)
ISBN 978-981-283-617-5

Volume 14: Solar Terrestrial (ST)
ISBN 978-981-283-619-9

Volume 15: Planetary Science (PS)
ISBN 978-981-283-621-2

A 6-Volume Set

Volume 16: Atmospheric Science (AS)
ISBN 978-981-283-809-4

Volume 17: Hydrological Science (HS)
ISBN 978-981-283-811-7

Volume 18: Ocean Science (OS)
ISBN 978-981-283-813-1

Volume 19: Planetary Science (PS)
ISBN 978-981-283-815-5

Volume 20: Solid Earth (SE)
ISBN 978-981-283-817-9

Volume 21: Solar Terrestrial (ST)
ISBN 978-981-283-819-3

A 6-Volume Set

Volume 22: Atmospheric Science (AS)
ISBN 978-981-4355-30-8

Volume 23: Hydrological Science (HS)
ISBN 978-981-4355-32-2

Volume 24: Ocean Science (OS)
ISBN 978-981-4355-34-6

Volume 25: Planetary Science (PS)
ISBN 978-981-4355-36-0

Volume 26: Solid Earth (SE)
ISBN 978-981-4355-38-4

Volume 27: Solar Terrestrial (ST)
ISBN 978-981-4355-40-7

A 4-Volume Set

Volume 28: Atmospheric Science (AS) & Ocean Science (OS)
ISBN 978-981-4405-67-6

Volume 29: Hydrological Science (HS)
ISBN 978-981-4405-70-6

Volume 30: Planetary Science (PS) and Solar & Terrestrial Science (ST)
ISBN 978-981-4405-73-7

Volume 31: Solid Earth Science (SE)
ISBN 978-981-4405-76-8

Advances in

Geosciences

Volume 31: Solid Earth Science (SE)

Editor-in-Chief

Kenji Satake

University of Tokyo, Japan

Volume Editor-in-Chief

Ching-Hua Lo

National Taiwan University, Taiwan

NEW JERSEY • LONDON • SINGAPORE • BEIJING • SHANGHAI • HONG KONG • TAIPEI • CHENNAI

Published by

World Scientific Publishing Co. Pte. Ltd.
5 Toh Tuck Link, Singapore 596224
USA office: 27 Warren Street, Suite 401-402, Hackensack, NJ 07601
UK office: 57 Shelton Street, Covent Garden, London WC2H 9HE

British Library Cataloguing-in-Publication Data
A catalogue record for this book is available from the British Library.

ADVANCES IN GEOSCIENCES
A 4-Volume Set
Volume 31: Solid Earth Science (SE)

ISBN 978-981-4405-66-9 (Set)
ISBN 978-981-4405-76-8 (Vol. 31)

Typeset by Stallion Press
Email: enquiries@stallionpress.com

Printed in Singapore.

EDITORS

REVIEWERS

The Editors of Volume 31 would like to acknowledge the following referees who have helped to review the manuscript publish in this volume:

Chien-Hsin Chang
Dayi Chen
Horng-Yue Chen
Kuo-En Ching
Hery Harjono
Nai-Chi Hsiao
Hsin-Hua Huang
P. S. Kulkarni
Kuan-Chuan Lin
Ting-Li Lin
Sanjay Patil
J. Bruce H. Shyu
Ramesh Pratap Singh
Yu Wang

PREFACE

The present volume set (volumes 28 to 31) of *Advances in Geosciences* (*ADGEO*) is the sixth round of *ADGEO* series edited by the Asia Oceania Geosciences Society (AOGS), and contains papers presented at the eighth annual meeting held in Taipei in 2011.

The AOGS is an international society legally registered in Singapore, aiming to cooperate and promote discussion on studies of the Earth and its environment, as well as the planetary and space sciences. To achieve this objective, the AOGS has held its annual meetings since 2004. The AOGS has six sections, Atmospheric Sciences (AS), Hydrological Sciences (HS), Ocean Sciences (OS), Planetary Sciences (PS), Solar and Terrestrial Sciences (ST), and Solid Earth Sciences (SE). In the current set, papers presented at AS and OS sections are included in volume 28, those at HS section are in volume 29, at PS and ST sections are in volume 30, and at SE section are in volume 31.

ADGEO is not a scientific journal, but a monograph series or proceeding volumes of the AOGS meetings. Only papers presented at the AOGS meetings are invited to *ADGEO* series, and are published after peer reviews. The first (volumes 1 to 5), second (6 to 9), third (10 to 15) sets corresponded to the second, third and fourth AOGS annual meetings. The fourth volume set (16 to 21) included papers presented at the fourth and fifth annual meetings, and the fifth set (22 through 27) included those at the sixth and seventh meetings.

As a young scientific society, AOGS needs to develop ways to promote information exchange and interaction among scientists in Asia and Oceania region, in the era of internet. Until we establish a journal or other means of publication, *ADGEO* is expected to serve as a publication tool among the AOGS members and society at large.

Finally, I would like to thank authors, reviewers, volume editors and volume editors-in-chief for their timely efforts to publish the current volume set, Meeting Matters International (MeetMatt) for developing and maintaining a system for submission, review and editorial processes, and World Scientific Publishing Company (WSPC) for the editorial, publication and marketing processes.

Kenji Satake
Editor-in-Chief

PREFACE TO SE VOLUME

The present volume successfully collects seven papers presented in the Solid Earth Section of the 2011 annual meeting of Asia Oceania Geosciences Society, held in Taipei. A glance through the table of contents shows the diversity of approaches to study methodological improvements in geochemistry, geophysics, geodetic and earthquake studies. The first group of papers discusses technical advances in geophysics, inducing application of particle swarm optimization techniques for locating earthquakes in Himalaya (Yadav *et al.*), limitation of earthquake early warning system in China (Ma *et al.*), development of the geodetic reference frame (APREF) in the Asia and Pacific region (Hu *et al.*), and application of VLBI and GNSS geodetic techniques in New Zealand (Takiguchi *et al.*). Paper by H. H. Aung interprets the relationship of the movement of Sagaing fault and historical earthquakes in Myanmar. The last two papers discuss the impacts of land subsidence and sea level rise to the coastal cities in Indonesia (Abidin *et al.*) and the physic-chemical characteristics of Quaternary sediments of the Terna River Basin in India and their potential sources (Babar *et al.*).

The editor would like to acknowledge all the conveners and contributors in Solid Earth Section of the meeting and is also indebted to the editors and reviewers whose efforts ensured the outstanding quality of these papers.

Ching-Hua Lo
SE Volume Editor-in-Chief

CONTENTS

Editors v

Reviewers vii

Preface ix

Preface to SE Volume xi

Metaheuristic Technique for Finding Earthquake Locations in NW Himalayan Region 1
A. Yadav, K. Deep, S. Kumar and R. Sushil

Challenging the Limit of EEW: A Scenario of EEWS Application Based on the Lessons of the 2008 Wenchuan Earthquake 11
X.-J. Ma, Z.-L. Wu, H.-S. Peng and T.-F. Ma

Towards the Densification of the International Terrestrial Reference Frame in the Asia and Pacific Region — Asia Pacific Reference Frame (APREF) 23
G. Hu, J. Dawson, M. Jia, M. Deo, R. Ruddick and G. Johnston

Towards Synergy of VLBI and GNSS Geodetic Techniques in Geologically Dynamic New Zealand 33
H. Takiguchi, S. Gulyaev, T. Natusch, S. Weston and L. Woodburn

Reinterpretation of Historical Earthquakes During 1929 to 1931, Myanmar 43
H. H. Aung

The Impacts of Coastal Subsidence and Sea Level Rise in Coastal City of Semarang (Indonesia) 59
H. Z. Abidin, H. Andreas, I. Gumilar, Y. Fukuda, S. L. Nurmaulia, E. Riawan, D. Murdohardono and Supriyadi

Physico-chemical Characteristics of Quaternary Sediments of Terna River Sub-basin East Central Maharashtra, India 77
Md. Babar, R. V. Chunchekar and B. B. Ghute

Advances in Geosciences
Vol. 31: Solid Earth Science (2011)
Eds. Ching-Hua Lo *et al.*

METAHEURISTIC TECHNIQUE FOR FINDING EARTHQUAKE LOCATIONS IN NW HIMALAYAN REGION

ANUPAM YADAV* and KUSUM DEEP†
Department of Mathematics
Indian Institute of Technology Roorkee,
Uttarakhand 24767, India
**anupuam@gmail.com*
†kusumfma@iitr.ernet.in

SUSHIL KUMAR
Wadia Institute of Himalayan Geology,
Dehradun, Uttarakhand 248001, India
sushil_rohella@yahoo.co.in

RAMA SUSHIL
Sri Guru Ram Rai Institute of Science and Technology,
Dehradun, Uttarakhand 248001, India
rmasushil@yahoo.com

In this article, we have used a new metaheuristic technique particle swarm optimization (PSO) for development of the earthquake location models. Two models of different crustal structures has been taken based on the arbitrary tardiness structure that defines the heterogeneity of the earth's crust. The problem of earthquake location is modeled as a least square function of travel times and solved by using the randomized search algorithm PSO. We have solved the problem by using an advanced version of PSOs. A real life data of earthquake in the NW Himalayan region has been taken for testing these developed models. The new locations are better than the existing results.

1. Introduction

Determination of hypocentral parameters: Epicentral latitude and longitude, depth and time of occurrence are the functions of arrival times of seismic P (and S) waves observed at a number of stations. Since arrival time observations may contain some errors, a minimum of three stations' data pertaining to each earthquake are required to obtain, in some prescribed sense, an optimized estimate of its hypocentral parameters provided that distribution of P-wave speed are known in that region.

A number of algorithms have been developed for estimating hypocentral parameters, from P-wave arrival times. These are broadly characterized in two categories depending upon whether the P-wave speed distribution is assumed or posed as unknowns to be determined by inverting P-wave arrival times at a larger number of recording stations.

The algorithm presented in this paper enables one to determine hypocentral parameters in a layered earth, in which two crustal models of the earth crust is taken to justify the availability of the particle swarm optimization (PSO) algorithm. Estimation of hypocentral parameters from arrival times of seismic waves, constitutes a geophysical inverse problem requires the solution to the corresponding forward problem, already available. The direct problem here is to obtain a formalism where P-wave arrival times from a given P-wave speed distribution is the region. This requires that the ray path between the known hypocenter and the recording station be traceable. We have chosen Clifford's travel time model; single layer as well as multilayer to calculate the travel time of seismic waves (P and S).

2. The Algorithm

2.1. *Model for single layered crustal structure*

Let V_0 be the velocity of P-wave. Then, the travel time for P-waves from source (x, y, z) to station $(x_0, y_0, 0)$ for a single layered crustal model[11] (See Fig. 1) is given by

$$t_s = \frac{[(x_0 - x)^2 + (y_0 - y)^2 + z^2]^{1/2}}{V_0}, \tag{1}$$

where t_s can be calculated by using the difference of arrival times of P and S waves and easily can be expressed as

$$t_{\text{pwave}} = \frac{V_s * \Delta t}{(V_p - V_s)}, \tag{2}$$

where V_s = speed of S-wave; V_p = speed of P-wave; Δt = travel time difference of P and S waves from source to station.

Now, we will design a least square function of t_{pwave} and t_s say f_s, by minimizing f the corresponding values of x, y and z will give the desired hypocenter.

$$f_s = \left[\sum_i (t_s - t_{\text{pwave}})^2 \right]^{1/2}, \tag{3}$$

where i = number of observation stations.

2.2. *Model for multilayered crustal structure*

Consider we will now crustal model of layer of velocity V_0 over a half space of velocity V_1. We must examine the cases of both direct and refracted rays. The refracted case again be treated analytically. Given a source at (x, y, z) and receiver $(x_0, y_0, 0)$, the travel time of P-waves can be expressed as (See Fig. 2)

$$t = \frac{\Delta}{V_1} + \frac{((V_1)^2 - (V_0)^2)^{1/2}}{V_0 * V_1} * (2D_0 - z), \tag{4}$$

where Δ = epicentral distance; D_0 = thickness of layer.

The implicit equation for travel time can be expressed as

$$t_m = \frac{[(x_1 - x_c)^2 + D_1^2]^{1/2}}{V_1} + \frac{[(x_c - x_0)^2 + D_0^2]^{1/2}}{V_0}. \tag{5}$$

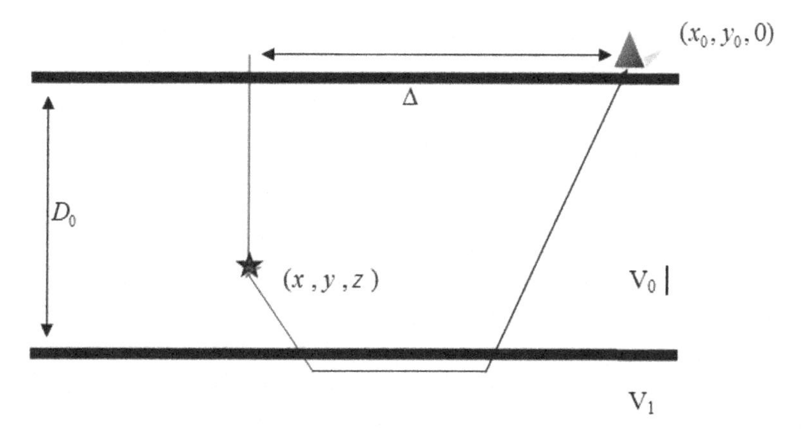

Fig. 1. Ray path diagram of critically refracted arrival.

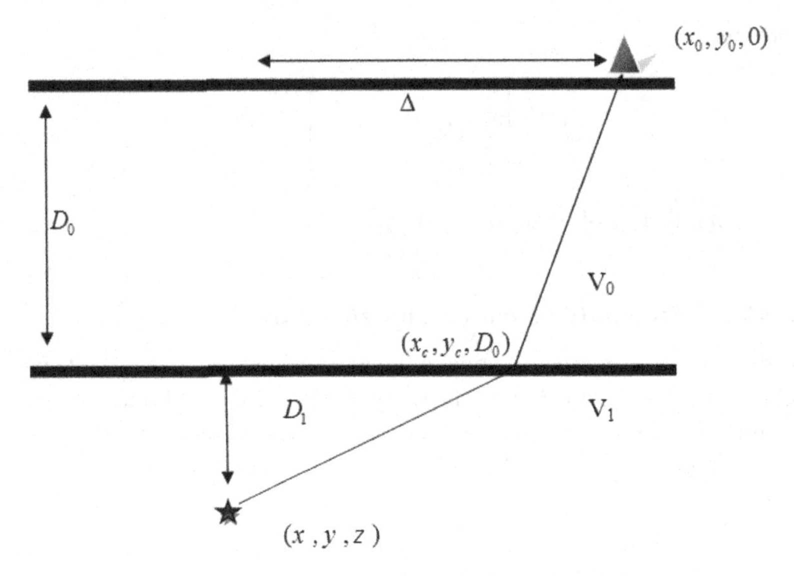

Fig. 2. Ray path diagram of multilayered structure.

Again the least square function for multilayered crustal structure is given by:

$$f_m = \left[\sum_i (t_m - t^i_{\text{pwave}})^2 \right]^{1/2}. \tag{6}$$

Now, our aim is to find the values of (x, y, z) for which f has the minimum functional value. We will find the minimum of f (corresponding f_s and f_m) by using one of the metaheuristic technique (PSO). The next section will give the brief introduction about particle swarm optimization technique and its applcation to this problem.

3. Particle Swarm Optimization

PSO,[3] which was first proposed by Kennedy and Eberhart in 1995,[3] is a population-based stochastic algorithm for continuous optimization. The algorithm is inspired by the social interaction behavior of bird's flocking and fish schooling. To search for the optimal solution, each individual, which is typically called a "particle" updates its flying velocity and current position iteratively according to its own flying experience and the other particles flying experience, By now, PSO has become one of the most popular optimization techniques for solving continuous optimization problems.

In original PSO, "S" particles cooperate to search for the global optimum in the n-dimensional search space. The ith ($i = 1, 2, \ldots, M$) particle maintains a position $X_i\,(x_i^1, x_i^2, \ldots, x_i^n)$ and velocity $V_i\,(v_i^1, v_i^2, \ldots, v_i^n)$. In each iteration, each particle uses its own search experience and the whole swarm's search experience to update the velocity and position. The updating rules are as follows:

$$v_i^j = v_i^j + c_1 r_1^i (pbest_i^j - x_i^j) + c_2 r_2^j (gbest^j - x_i^j), \tag{7}$$

$$x_i^j = x_i^j + v_i^j, \tag{8}$$

where $Pbest_i\,(pbest_i^1, pbest_i^2, \ldots, pbest_i^n)$ is the best solution yielded by the ith particle and $Gbest\,(gbest^1, gbest^2, \ldots, gbest^n)$ is the best-so-far solution obtained by the whole swarm. c_1 and c_2 are two parameters to weigh the importance of self-cognitive and social-influence, respectively. r_1^j and r_2^j are random numbers uniformly distributed in $[0, 1]$, and $j\,(j = 1, 2, 3, \ldots, n)$ represents the ith dimension.

Different varieties of PSO variants have been proposed till date from the very first appearance of its original version. For solving the current problem of earthquake location, we have chosen the newly developed one of the variant of original PSO, LXPSO.[1] The next section will give the brief idea about LXPSO.

3.1. *Lapalce's crossover PSO (LXPSO)*

This version of PSO has been proposed by Bansal *et al.*[1] Herein, PSO is hybridized by incorporating Laplace's crossover (Earlier designed for genetic algorithm by Deep and Thakur, 2007). In LXPSO algorithm, the author has used a term Laplacian particle, and the description of the same is discussed in the following paragraph.

Laplace's crossover proposed by Deep and Thakur[6] is based on the Laplace's distribution. This parent centric operator is called Laplace's operator (LX), here two offsprings $y_1\,(y_1^1, y_1^2, \ldots, y_1^n)$ and $y_2\,(y_2^1, y_2^2, \ldots, y_2^n)$ are generated from a pair of parents $x_1\,(x_1^1, x_1^2, \ldots, x_1^n)$ and $x_1\,(x_2^1, x_2^2, \ldots, x_2^n)$ using LX. First, a uniformly distributed random number $u_d \in (0, 1)$ is generated, then from Laplace's distribution function, the ordinate β_d is calculated so that the area under probability curve excluding area a (location parameter) to β_d is equal to the chosen random number u_d.

$$\beta_d = \begin{cases} a - b\,\log_e(1 - 2u_d), & u_d \leq \frac{1}{2} \\ a - b\,\log_e(2u_d - 1), & u_d > \frac{1}{2}. \end{cases}$$

Here, b is called the scale parameter. The offspring are then given by the equations:

$$y_{1D} = x_{1d} + \beta_d \mid (x_{1d} - x_{2d}) \mid d = 1, 2, 3, \ldots, n,$$

$$y_{2D} = x_{2d} + \beta_d \mid (x_{1d} - x_{2d}) \mid d = 1, 2, 3, \ldots, n.$$

Based on the Laplacian operator as described above, two particles are formed. The best particle (in terms of fitness) is selected. This new particle is called Laplacian particle. Algorithm of LXPSO is as follows:

Create and Initialize a n-dimensional swarm S;
For t = 1 to the maximum_iterations,
For m = 1 to S,
For d = 1 to n,
Apply the velocity update equation (6) and update the position using equation (7)
End for d;
Compute fitness of updated position;
If needed, update historical information for Pbest and Gbest;
End-for-n;
Select two random particles from the current swarm for interaction. Generate the Lapalcian particle as a result of this interaction. Replace the worst particle in the swarm with Lapalcian particle;
Compute fitness of Lapalcian Particle;
If needed, update historical information for P_{best} *and* G_{best}*;*
Terminate if meets problem requirements;
End for t;

3.2. Selection of parameters for LXPSO

Dimension (n) of the search space is three, i.e. $n = 3$. The decision variables in algorithm are latitude, longitude and the depth of the hypocenter. The selection of parameter is done according to Kennedy and Clerc.[4] The swarm size is $13[10 + (\text{int})(2 * \text{sqrt}(D)]$, $c_1 = c_2 = 0.5 + \log(2)$. Number of iterations are 200. The stopping criterion for the algorithm is designed two-way jointly, where the first one is the number of iterations and the second one is the minimum error for this (the minimum defined error is 0.0001). Our algorithm will execute whichever condition is met first.

P-wave speed distribution has been taken from Sushil Kumar *et al.*[9] The speed distribution of P waves in the Hindu Kush and Garhwal regions is $V_0 = 6.2\,\mathrm{Km/sec}$ and $V_1 = 8.37\,\mathrm{Km/sec}$, respectively.

4. Experiments

The problem is solved using LXPSO algorithm. Earthquake data has been taken from Wadia Institute of Himalayan Geology Dehradun. Eight stations are used to observe the data and the problem is solved for single layer as well as multilayered hypocentral location.To test the method and algorithm, we have taken an event of earthquake which is observed at eight stations established in the state of Uttarakhand, India. Locations of these stations are in Table 1.

4.1. *Single layered crustal model*

First, we will evaluate the experimental results of hypocentral location with single layered model.[11] An event of earthquake was recorded on 14 August 2008 at Wadia Institute of Himalayan Geology Deh radun whose hypocentral parameters are 31.741°N, 70.832°E, 10.0 Km. The difference between travel times of P and S waves observed at the stations listed in Table 2.

Table 1. Location of observation stations in Uttarakhand, India.

Station name	Latitude (Degree)	Longitude (Degree)
Gaurikund	30.65 N	79.02 E
Adibadri	30.14 N	79.20 E
Kharsali	30.97 N	78.43 E
Chakrata	30.71 N	77.86 E
Kotkhai	31.10 N	77.58 E
Nahan	30.54 N	77.25 E
Deoband	29.71 N	77.73 E
Tapovan	30.49 N	79.61 E
Ghuttu	30.53 N	78.74 E
Dehradun	30.33 N	78.10 E

Table 2. Location of observation stations in Uttarakhand, India.

Station name	Difference in travel time of P and S wave Δt (sec)
Gaurikund	18.78365
Adibadri	17.23289
Kharsali	16.03074
Chakrata	15.52536
Kotkhai	14.77106
Nahan	15.81494
Deoband	19.63537
Tapovan	17.83978
Ghuttu	16.52489
Dehradun	15.52540

Table 3. Location of observation stations in Uttarakhand, India.

Latitude	Longitude	Depth	Function value (f)
30.33049 N	70.07944 E	15.35585 Kms	0.000122
30.01800 N	70.48918 E	15.01604 Kms	0.000119
30.42796 N	70.48241 E	15.13957 Kms	0.000139
30.47073 N	70.24463 E	15.02327 Kms	0.000140
30.34208 N	70.40334 E	15.04895 Kms	0.000109
30.38190 N	70.07151 E	15.41502 Kms	0.000125
30.37454 N	70.21257 E	15.35019 Kms	0.000122
30.19768 N	70.46153 E	15.15982 Kms	0.000133
30.33036 N	70.39927 E	15.47891 Kms	0.000136
30.08628 N	70.48358 E	15.01736 Kms	0.000138

Note: Result of 10 runs.

We have minimized the function f and evaluated the corresponding values of x, y and z as a resulting hypocenter. Table 3 shows the results.

Table 3 shows the earthquake location which was observed by eight stations and is the result of 10 runs with function value f. From Table 3, it can be observed that the estimated hypocentral parameters are pretty close to the location estimated at Wadia Institue of Himalayan Geology.

4.2. *Multilayered crustal model*

In this section, we have taken another event of earthquake recorded on 16 December 2008 at WIHG with greater depth, the recorded location is 30.039°N, 78.856°E, 94.4 Km. The travel time difference of P and S waves for this event observed at stations listed in Table 2 are in Table 4.

Table 4. Location of observation stations in Uttarakhand, India.

Station name	Difference in travel time of P and S wave Δt (sec)
Gaurikund	39.8751
Adibadri	40.4200
Kharsali	35.7599
Chakrata	40.4654
Kotkhai	38.7839
Nahan	35.5837
Deoband	36.6665
Tapovan	38.2724
Ghuttu	39.8751
Dehradun	40.4200

Table 5. Location of observation stations in Uttarakhand, India.

Latitude	Longitude	Depth	Function value (f)
77.7535	28.6734	110.1864	0.0007895
77.8778	28.0337	110.2654	0.0008740
77.1237	28.2855	110.2757	0.0008743
77.2149	28.2409	110.0648	0.0008716
77.0470	28.1099	110.1698	0.0008739
77.2325	28.3295	110.1545	0.0008737
76.9987	28.0119	110.2241	0.0008709
77.2230	28.0960	110.1521	0.0008707
77.2504	28.0160	110.1323	0.0008728

Note: Result of 10 runs.

Now we have minimized the function f with one additional variable x_c. The corresponding values of x, y, z and x_c for which f is minimum will give the best possible result of hypocentral location. Table 5 gives the minimized value of f with corresponding values of x, y and z.

5. Discussion and Conclusion

Two earthquake events are taken with relevant observations at eight stations. Hypocentral parameters are evaluated with the discussed model for single layered as well as multilayered crustal structure by using LXPSO algorithm. The results obtained by the LXPSO algorithm are very promising and quite matching with the estimated hypocentral parameters at WIHG. This article is a very good application of metaheuristic techniques in seismological problems and can be applied to the other seismological problems where optimization is needed.

Acknowledgments

Authors are thankful to Council of Scientific and Industrial Research, Human Resource and Development Group, Govt. of India for funding this work. We are also thankful to the Institute Computer Center, Indian Institute of Technology Roorkee and Wadia Institute of Himalayan Geology Dehradun for providing necessary computational facilities and earthquake data.

References

1. J. C. Bansal, K. Veeramachaneni, Deep Kusum and L. Osadciw, Information sharing strategy among particles in particle swarm optimization using laplacian operator, *IEEE Swarm Intelligence Symposium*, Vol. 30 (2009), pp. 30–36.
2. K. Shanker, C. Mohan and K. N. Khattri, *Tectonophysics* **198**(1) (1991) 73–80.
3. J. Kennedy and R. Eberhart, Particle swarm optimization, *Neural Networks, Proceedings, IEEE International Conference*, Vol. 4 (1995), pp. 1942–1948.
4. M. Clerc and J. Kennedy, Standard PSO. Available at http://www.particleswarm.info/Standard_PSO_2006.c.
5. K. L. Kaila, V. G. Krishna and N. Hari, *Bull. Seismol. Soc. Am.* **59**(5) (1969) 1949–1967.
6. K. Deep and M. Thakur, *Appl. Math. Comput.* **188**(1) (2007) 895–911.
7. M. M. Matveyeva and A. A. Lukk, *Geophys. J. Roy. Astron. Soc.* **49** (1977) 87–114.
8. A. Ram Mereu, R. F., *Geophys. J. Roy. Astron. Soc.* **49** (1977) 87–113.
9. S. Kumar and T. Sato, *J. Himalayan Geology* **24**(2) (2003) 77–85.
10. H. Dong-xue and W. Gai-yun, Application of particle swarm optimization to seismic location, *Third Int. Conf. of Genetic and Evolutionary Computing* (2009), pp. 641–644.
11. C. H. Thurber, *Bull. Seismol. Soc. Am.* **75**(3) (1985) 779–790.
12. I. Sarkar, R. Chander, K. N. Khattri and V. K. Gaur, Estimations of hypocentral parameters of local earthquakes when crustal layers have constant P-velocities and dipping interfaces, *Proc. Indian Acad. Sci Earth Planet. Sci.*, Vol. 96, No. 3 (1987), pp. 229–238.
13. C. Wei-Neng, J. Zhang, H. S. H. Chung, Z. Wen-Liang, W. Wei-Gang and Yu-Hui Shi, *IEEE Transactions on Evolutionary Computation* **14**(2) (2010) 278–299.

Advances in Geosciences
Vol. 31: Solid Earth Science (2011)
Eds. Ching-Hua Lo *et al.*

CHALLENGING THE LIMIT OF EEW: A SCENARIO OF EEWS APPLICATION BASED ON THE LESSONS OF THE 2008 WENCHUAN EARTHQUAKE

XIAOJING MA
Institute of Geophysics, China Earthquake Administration, Beijing 100081, China

ZHONGLIANG WU*
Laboratory of Computational Geodynamics, Graduate University of the Chinese Academy of Sciences, Beijing 100049, China
Institute of Geophysics, China Earthquake Administration, Beijing 100081, China
wuzhl@gucas.ac.cn

HANSHU PENG
Institute of Geophysics, China Earthquake Administration, Beijing 100081, China

TENGFEI MA
Institute of Geophysics, China Earthquake Administration, Beijing 100081, China

Based on the lessons and experiences of the 2008 Wenchuan earthquake, we discussed a scenario of a special type of earthquake early warning system (EEWS), in which the rupture extent is larger than the "blind zone" of EEWS. We used the strong motion recordings and seismic recordings of the Wenchuan earthquake for determining the first-3-seconds EEW magnitude/s and for the analysis of the "EEW schedule". It is shown that near-source recordings are able to provide the first-3-seconds τ_c magnitude, while the τ_c magnitude tends to be smaller for stations near to the epicenter; *Pd* magnitude tends to underestimate the earthquake. Significantly useful warning time would be available, if an EEWS were deployed on site.

*Corresponding author.

1. Introduction

Recently, with the development of real-time seismology,[7,8] various types of earthquake early warning systems (EEWS) have been developed in many areas such as Mexico City,[5] Japan,[10] Taiwan,[6,20–24,26] Istanbul,[1,13] Italy,[29] California,[2,3,25,28] and the Beijing Capital Region,[14] either experimentally or for actual operation. In the development of EEWS, one of the critical comments often met is that because of the "blind zone", in which differential time between P and S arrivals does not permit a practical early-warning, the EEWS is actually of little use in saving life and property, questioning the values of EEWS in the reduction of earthquake disasters. Indeed, this comment is to much extent reasonable, since the "blind zone" of EEWS often experiences high intensity shaking.

However, for inland great earthquakes, with the extended rupture larger than the "blind zone" of EEWS, there would be another important opportunity to save life and property if an EEWS were deployed and used properly, as shown by the scenario case in this article.

In this article, we investigate a special type of EEWS. We discuss this problem by a scenario of a great inland earthquake with an extended rupture, larger than the "blind zone" which is typically about 50 km. Actually, this is the case of the 2008 Wenchuan earthquake with its rupture zone being more than five times of the "blind zone".

In a conceptual perspective, this is by no means a new idea. In the personal communications of Chinese seismologists (Mengtan Gao and Xing Jin, internal presentation), concepts similar to this scenario have been proposed and discussed based on the experiences and lessons of evacuation during the Wenchuan earthquake, especially in Shaanxi and Gansu provinces to the north of the Wenchuan earthquake rupture. The case that the station is near or even "above" the earthquake rupture has also been discussed by Shieh *et al.*[16] For the Wenchuan earthquake, retrospective discussion on the early warning scenario was conducted by Wang *et al.*[19] and Wan *et al.*,[17] shortly after the earthquake.

In this paper, we discuss two practical issues: (1) whether the seismic and/or strong motion recordings of the first-3-seconds are able to provide a correct or "reasonable" estimate of the τ_c (average period)/Pd (peak amplitude of the filtered vertical displacement) magnitude of the mainshock; (2) whether there were enough time for the early warning message to be sent to the locations within the rupture zone.

2. Data Used and Method for the EEW Magnitude Estimation

We used seismic recordings, deployed for the monitoring of earthquakes, and strong motion recordings, deployed mainly for earthquake engineering studies, provided by the Earthquake Administration of Sichuan Province and the China Strong Motion Networks Center, respectively. The 25 strong motion stations (recording broadband ground accelerations, with sampling rate 200 sps except the AXT station whose sampling rate is 500 sps) and 10 seismic stations (recording broadband ground velocities, with sampling rate 100 sps) distribute along the mid-and-northern Longmenshan fault zone, which ruptured during the Wenchuan earthquake. The seismic stations use the CMG-3ESPC instruments with the velocity frequency response flat for 60 s–50 Hz, except the CD2 station which uses JCZ-1 instrument with the velocity frequency response flat for 360 s–20 Hz. For the description of the instrumental and site characteristics of the strong motion stations and the strong motion recordings of the 2008 Wenchuan earthquake, see Li *et al.*[9] Figure 1 shows the distribution of stations used for the analysis. For reference, Fig. 1 also shows the slip distribution of the Wenchuan mainshock inverted from regional and teleseismic recordings.[18]

For the Wenchuan mainshock, most of the seismic recordings clipped when the strong S-waves arrived. In our approach, vertical component recordings were used for the analysis. In the pre-processing of the data, as a convention,[19] a 0.075 Hz high-pass filter was used to the integrated data.

For the estimation of magnitude using the first-3-seconds of the P arrivals, we adopted the widely used and proven useful τ_c and Pd methods. The τ_c method, firstly proposed by Nakamura[10] and developed by Allen and Kanamori[2] and Kanamori,[7] used the ground velocity $\dot{u}(t)$ and the ground displacement $u(t)$ to estimate the magnitude via

$$\tau_c = 2\pi/\sqrt{r} \tag{1}$$

with

$$r = \frac{\int_0^{\tau_0} \dot{u}^2(t)dt}{\int_0^{\tau_0} u^2(t)dt}. \tag{2}$$

We used the scaling relation between magnitude and τ_c for Southern California,[25] Japan,[17] Taiwan[27] and Wenchuan,[19] respectively, in estimating the magnitude. We also used the Pd method, an estimation of magnitude

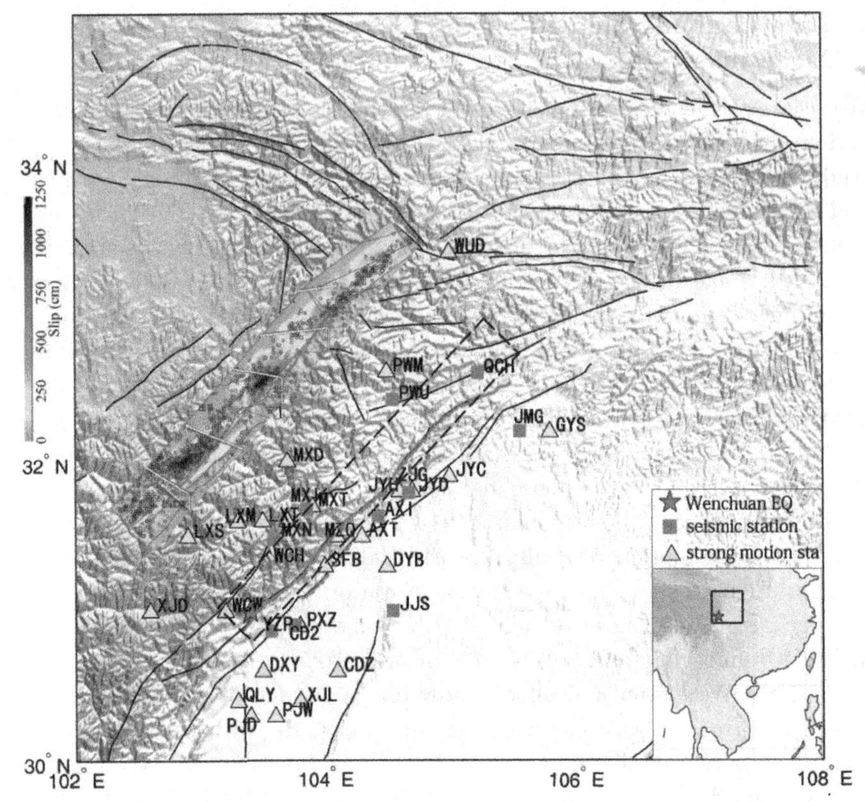

Fig. 1. Distribution of seismic stations (red squares) and strong motion stations (yellow triangles) used in this study. The red lines approx the surface rupture during the earthquake. The projection of seismic slip on the fault, corresponding to the dashed box, is from Wang *et al.*[18] with the color bar indicating the amount of the slip. Position of the study region is marked by the indexing figure to the bottom right.

using the peak amplitude of displacement within the first-3-seconds after the P-arrival. For comparison, we used the magnitude – (*Pd*, hypocentral distance) relation for Wenchuan,[19] with the reference of the scaling relations for other regions.

3. Magnitude Estimation

Necessity of the present investigation lies in the fact that the mainshock rupture of the Wenchuan earthquake lasted several tens of seconds, while the first-3-seconds of P-arrival only records a very small portion of

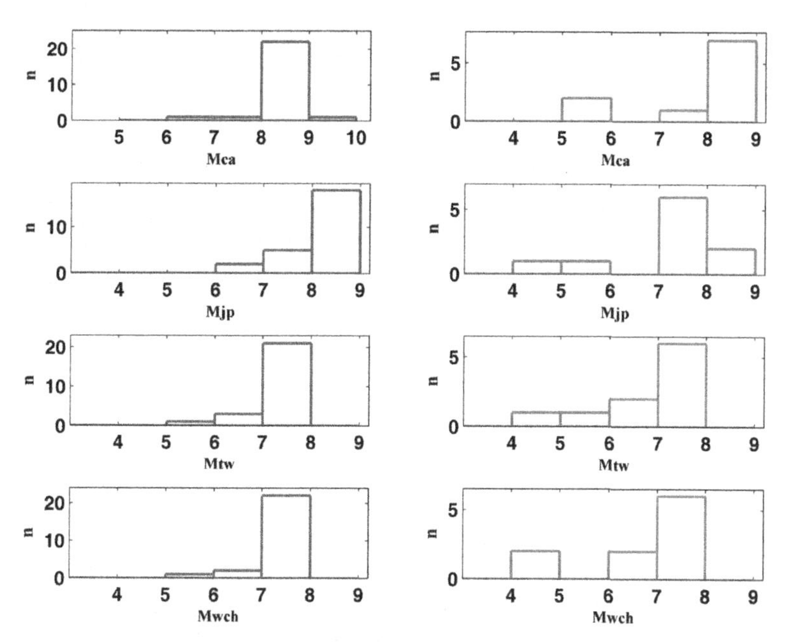

Fig. 2. Estimated τ_c magnitude, using the empirical scaling relation between τ_c and magnitude for Southern California,[25] Japan,[17] Taiwan[27] and Wenchuan,[19] respectively. The blue histograms to the left represent the results from strong motion records, and the red ones to the right represent the results from seismic records.

the earthquake rupture. Whether this onset information is sufficient for estimating the final magnitude of the earthquake is the first problem which has to be answered by real data. Considering the comparison of the first-3-seconds P arrivals and the size of the whole rupture, the Wenchuan earthquake also provides an interesting case for investigating the controversial argument on whether earthquake rupture is deterministic.[11,12,15]

In Fig. 2, the estimated τ_c magnitudes are displayed using the empirical τ_c–magnitude relation of four different areas. It can be seen that majority of the results are distributed around magnitude 8, while the results using relations for California and Japan are relatively larger on the whole. The calculated magnitudes from seismic records are apparently more dispersive compared to those from strong motion records. Contributing to such a difference, the WCH and the YZP station, located near to the epicenter, provide extremely low estimates.

Despite the scattering of the estimated magnitudes, it is noticeable that the τ_c magnitudes using different scaling relations generally well reflect the size of the mainshock.

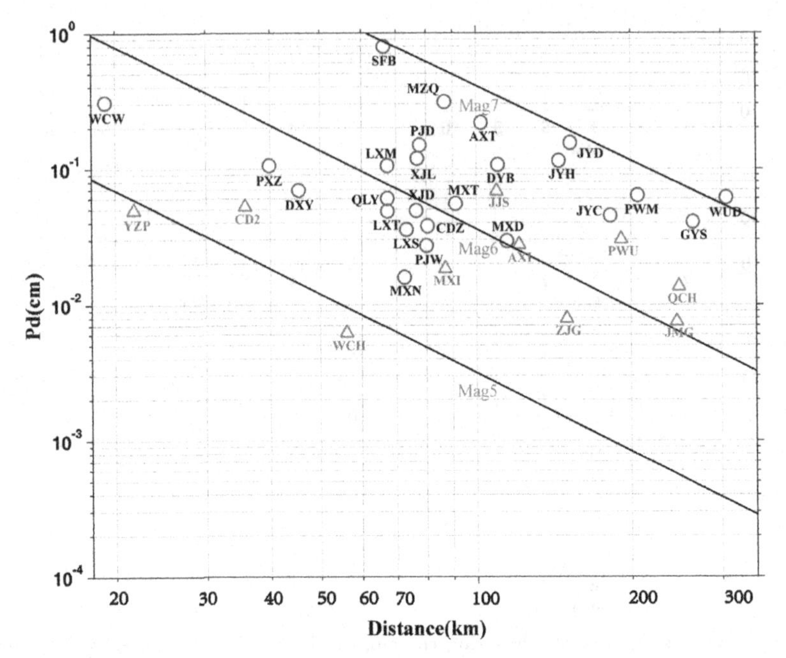

Fig. 3. Estimated *Pd* magnitudes using records from strong motion stations (blue circles) and seismic stations (red triangles), with the names of the stations labeled to the data points. The straight lines represent the empirical distance-dependent *Pd*–magnitude relation, standing for magnitude 5, 6 and 7, respectively.[19]

The estimated *Pd* magnitude from strong motion records (blue circles) and seismic records (red triangles) are displayed in Fig. 3. The empirical distance-dependent *Pd*–magnitude relation for Wenchuan[19] is shown in the figure for reference. The *Pd* magnitudes scatter between five and seven, significantly underestimate the mainshock. Another feature is that the calculated magnitudes from seismic records are about 0.5 magnitude units smaller compared to those from strong motion records. Because the seismic stations and the strong motion stations are not colocated, the comparison using a limited number of stations cannot arrive at more geophysical conclusions.

For the τ_c values of 12 records within epicenter-distance 75 km, five of them are less than 3.0, and seven are less than 4.0, corresponding to (an underestimated) magnitude 6.8 and 7.2, respectively (Fig. 4). The six stations with anomalously small τ_c values are highlighted blue in Fig. 5. It turns out that records from stations close to the epicenter might result in small τ_c values. Reason for this might be that near to the nucleation point of great earthquakes, there are more high-frequency seismic waves radiated,

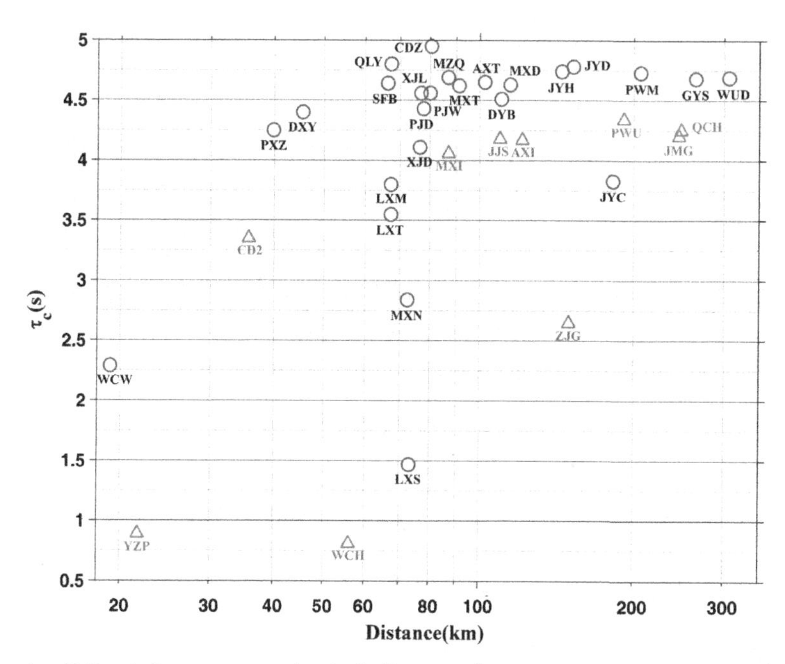

Fig. 4. Estimated τ_c versus epicentral distance, from strong motion stations (blue circles) and seismic stations (red triangles), with the names of the stations labeled to the data points.

which attenuate fast with the increase of the epicenter distance. The high-frequency contents of seismic waves raise the dominant frequency of the ground motion, and thus reduce the τ_c values.

4. Time Schedule for the Early Warning

In considering the EEWS facing to inland great earthquakes such as the Wenchuan case, one has to consider the complexity caused by the mixed effect of the finite propagation of seismic waves and the finite propagation of seismic rupture. In this case, the traditional concept of front-detection early warning and on-site early warning might be mixed together.

We take six strong motion stations distributing along the fault to illustrate the schedule for the early warning practice. These stations are shown in Fig. 5 by the names of stations in red. Figure 6 shows the vertical component waveforms of the six strong motion stations aligned according to their distances to the epicenter. The waveforms are synchronized by starting 3 sec before the onset of the first P arrival. From the figure the whole

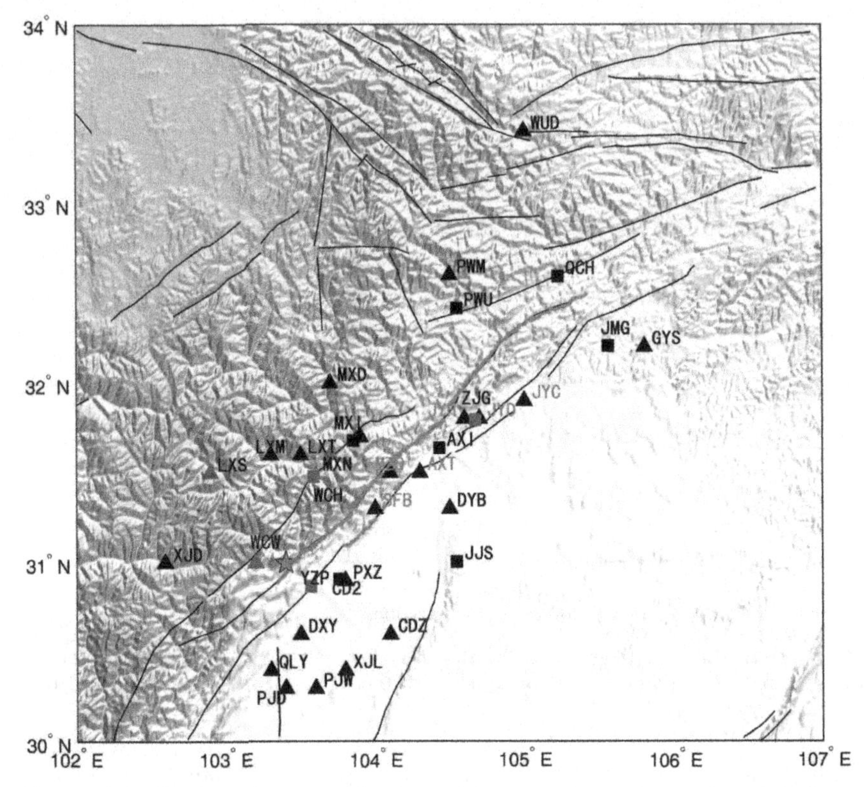

Fig. 5. Distribution of stations. Stations with extremely small τ_c values (<3.0) are highlighted in blue. Except station ZJG, all these stations are located near the epicenter. Stations aligned along the rupture fault (to be used in Fig. 6) are labeled with red station names.

process of the seismic strong motion can be seen. For the stations with epicenter distances larger than 50 km, there are some 19 sec (SFB) to 54 sec (JYC) from the P-arrivals to the peak S waves, providing "on-site early warning time" from 14 sec (SFB) to 49 sec (JYC). Note that positions from SFB to JYC are all above the rupture of the mainshock. In the Wenchuan earthquake, some local people living in the north part of the Longmenshan fault zone used this time for the evacuation (Mengtan Gao, 2009, personal communication). If warning message could be sent from the near-epicenter stations to the faraway stations, for example, from SFB to MZQ, AXT, ..., and JYC, then some more 5 to 20 sec "front-detection early warning time" can be obtained for JYC. Therefore, even for the regions "just above" the earthquake rupture, useful "on-site early warning time" — the time from the

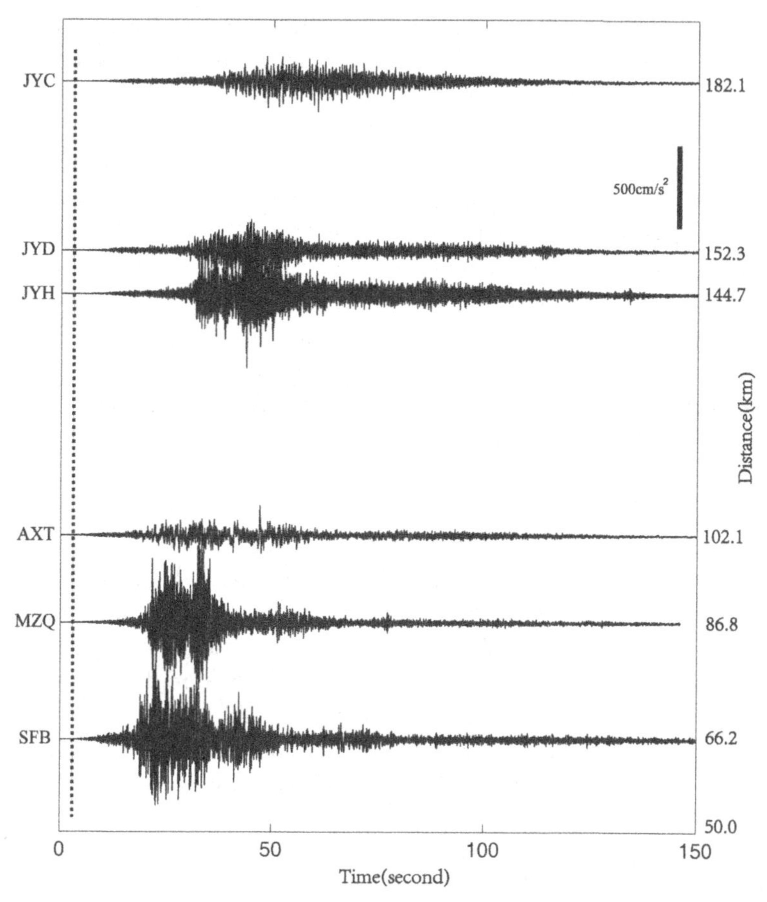

Fig. 6. Vertical component waveforms of the six strong motion records which are aligned along the rupture fault, listed according to their distances to the epicenter. Each waveform starts 3 sec before the onset of the first P arrival.

first P-arrivals to the peak S-waves — is available. If nearby stations could send the message to the faraway stations, then some more "front detection early warning time" could be available. But for the front detection early warning, cautions has to be taken due to the problem of underestimation, as discussed in the last section.

5. Discussions and Conclusions

Using the strong motion data and seismic data of the Wenchuan earthquake, we conducted a retrospective case study on a scenario of EEWS. We show that for the great inland earthquakes with rupture zones larger than the "blind zone" of EEWS, much more could be done for saving life and property by the proper design and application of EEWS. Quantitatively, we used real observational data in the analysis and confirmed that, for the case of a great inland earthquake with an extended rupture: (1) Near source recordings are able to provide the first-3-seconds τ_c magnitude, while the τ_c magnitude tends to be smaller for stations near to the epicenter; (2) *Pd* magnitude tends to underestimate the earthquake; (3) Significantly useful warning time is available, and in this case on-site EEW and front-detection EEW have mixed together; because of the underestimation by nearby stations, the "on-site early warning mechanism" is more effective. These features can be used in the design of the EEWS facing to both "regular earthquakes" and the rare great inland earthquakes.

The Wenchuan-type inland major to great earthquakes are not rare cases, at least for the Chinese mainland, in which several earthquakes with magnitude larger than 7.5 have occurred since 1997 (including the M_S7.9 Mani, Tibet earthquake in 1997; and the M_S8.1 Kokoxili, Qinghai earthquake in 2001 — fortunately, neither of them located in populated areas). In all these earthquake cases, the rupture zone was significantly larger than the "blind zone" of EEWS. If an EEWS were deployed before the earthquake, then significant warning time could be obtained even in some regions near or "just above" the seismic rupture. In the tectonic view, Chinese continent can be divided into several tectonic blocks, with most of the major, and all of the great earthquakes located near their boundary zones.[4] This means that the EEWS facing to inland great earthquakes is to much extent a practical issue. The discussion on the Wenchuan scenario might be useful in future, not only for continental China but also for other regions with inland earthquake risk.

Acknowledgments

Thanks to the Earthquake Administration of Sichuan Province and the China Strong Motion Networks Center for providing the seismic and strong motion data, and to Profs. Mengtan Gao and Xing Jin for stimulating discussion. Discussion with Prof. Yih-Min Wu helped much in the EEW magnitude estimation. Drs. Changsheng Jiang, Feng Long, Yong Zhang and staff of the Data Management Center of the China National Seismograph Network helped in collecting the data. This work is supported by the WFSD project.

References

1. H. Alcik, O. Ozel, N. Apaydin and M. Erdik, *Geophys. Res. Lett.* **36** (2009) L00B05, doi:10.1029/2008gl036659.
2. R. M. Allen and H. Kanamori, *Science* **300** (2003) 786–789, doi:10.1126/science.1080912.
3. R. M. Allen, *Seism. Res. Lett.* **77** (2006) 371–376, doi:10.1785/gssrl.77.3.371.
4. Y. Chen and D. C. Booth, *The Wenchuan Earthquake of 2008: Anatomy of a Disaster* (Beijing, Science Press, 2011).
5. J. E. Aranda, A. Jimenez, G. Ibarrola, F. Alcantar, A. Aguilar, M. Inostroza and S. Maldonado, *Seism. Res. Lett.* **66** (1995) 42–53.
6. N.-C. Hsiao, Y.-M. Wu, T.-C. Shin, L. Zhao and T.-L. Teng, *Geophys. Res. Lett.* **36** (2009) L00B02, doi:10.1029/2008gl036596.
7. H. Kanamori, *Annu. Rev. Earth Planet. Sci.* **33** (2005) 195–214, doi:10.1146/annurev.earth.33.092203.122626.
8. H. Kanamori, E. Hauksson and T. Heaton, *Nature* **390** (1997) 461–464.
9. X. Li, Z. Zhou, M. Huang, R. Wen, H. Yu, D. Lu, Y. Zhou and J. Cui, *Seism. Res. Lett.* **79** (2008) 844–854, doi:10.1785/gssrl.79.6.844.
10. Y. Nakamura, On the urgent earthquake detection and alarm system (UrEDAS), *Proceedings of the Ninth World Conference on Earthquake Engineering* **VII** (Balkema, Rotterdam, Netherlands, 1988) 673–678.
11. E. L. Olson and R. M. Allen, *Nature* **442** (2006), doi:10.1038/nature04964.
12. E. L. Olson and R. M. Allen, *Nature* **438** (2005) 212–215, doi:10.1038/nature04214.
13. A. Oth, M. Böse, F. Wenzel, N. Köhler and M. Erdik, *J. Geophys. Res.* **115** (2010) B10311, doi:10.1029/2010jb007447.
14. H. Peng, Z. Wu, Y.-M. Wu, S. Yu, D. Zhang and W. Huang, *Seism. Res. Lett.* **82** (2011) 394–403, doi:10.1785/gssrl.82.3.394.
15. P. Rydelek and S. Horiuchi, *Nature* **442** (2006), doi:10.1038/nature04963.
16. J.-T. Shieh, Y.-M. Wu, L. Zhao, W.-A. Chao and C.-F. Wu, *Soil. Dyn. Earthquake Eng.* **31** (2011) 240–246, doi:10.1016/j.soildyn.2009.12.015.
17. K. Wan, S. Ni, X. Zeng and P. Sommerville, *Sci. China Ser. D.* **52** (2009) 155–165, doi:10.1007/s11430-009-0007-1.
18. W. M. Wang, L. F. Zhao, J. Li and Z. X. Yao, *Chinese J. Geophys.* (in Chinese), **51** (2008) 1403–1410.
19. W. Wang, S. Ni, Y. Chen and H. Kanamori, *Geophys. Res. Lett.* **36** (2009) L16305, doi:10.1029/2009gl038678.
20. Y.-M. Wu and T.-L. Teng, *Bull. Seismol. Soc. Am.* **92** (2002) 2008–2018, doi:10.1785/0120010217.
21. Y.-M. Wu, *Seism. Res. Lett.* **71** (2000) 338–343, doi:10.1785/gssrl.71.3.338.
22. Y.-M. Wu, *Bull. Seismol. Soc. Am.* **95** (2005a) 347–353, doi:10.1785/0120040097.
23. Y.-M. Wu, *Bull. Seismol. Soc. Am.* **95** (2005b) 1181–1185, doi:10.1785/0120040193.
24. Y.-M. Wu, J. Chung, T. Shin, N. Hsiao, Y. Tsai, W. Lee and T. Teng, *Seism. Res. Lett.* **68** (1997) 931–943, doi:10.1785/gssrl.68.6.931.

25. Y.-M. Wu, H. Kanamori, R. M. Allen and E. Hauksson, *Geophys. J. Int.* **170** (2007) 711–717, doi:10.1111/j.1365-246X.2007.03430.x.
26. Y.-M. Wu, T.-C. Shin and Y.-B. Tsai, *Bull. Seismol. Soc. Am.* **88** (1998) 1254–1259.
27. Y.-M. Wu, H.-Y. Yen, L. Zhao, B.-S. Huang and W.-T. Liang, *Geophys. Res. Lett.* **33** (2006) L05306, doi:10.1029/2005gl025395.
28. G. Wurman, R. M. Allen and P. Lombard, *J. Geophys. Res.* **112** (2007) B08311, doi:10.1029/2006jb004830.
29. A. Zollo, G. Iannaccone, M. Lancieri, L. Cantore, V. Convertito, A. Emolo, G. Festa, F. Gallovič, M. Vassallo, C. Martino, C. Satriano and P. Gasparini, *Geophys. Res. Lett.* **36** (2009) L00B07, doi:10.1029/2008gl036689.

Advances in Geosciences
Vol. 31: Solid Earth Science (2011)
Eds. Ching-Hua Lo *et al.*

TOWARDS THE DENSIFICATION OF THE INTERNATIONAL TERRESTRIAL REFERENCE FRAME IN THE ASIA AND PACIFIC REGION — ASIA PACIFIC REFERENCE FRAME (APREF)*

G. HU[†], J. DAWSON, M. JIA, M. DEO, R. RUDDICK and G. JOHNSTON
Geoscience Australia, Cnr Jerrabomberra Ave
and Hindmarsh Dr Symonston Canberra
ACT 2601, Australia
[†]*guorong.hu@ga.gov.au*

Asia Pacific Reference Frame (APREF) is a joint initiative of the Permanent Committee on GIS Infrastructure for Asia and the Pacific (PCGIAP) and the International Association of Geodesy (IAG) with the objective of further developing the regional geodetic framework of the region. Currently, APREF incorporates GNSS data from a Continuously Operating Reference Station (CORS) network of approximately 400 stations, contributed by 28 countries, which are processed by three Local Analysis Centers (LACs). The contributions of the LACs are combined into daily, weekly and multi-year cumulative solutions, aligned to International Terrestrial Reference Frame (ITRF), in SINEX format using the CATREF software. These solutions provide reliable access to the ITRF while also enabling quality assessment of the performance of participating APREF stations. In this paper, two case studies of the applications of the APREF products are presented: The detection of co-seismic displacement and monitoring the stability of CORS.

1. Introduction

The broad objectives of APREF are to improve the definition, realization and maintenance of the regional geodetic framework in support of both scientific and regional geospatial activities by densifying the ITRF in the Asia and Pacific region. APREF is a collaboration of the Geodetic Technologies and Applications Working Group of the PCGIAP, which operates under the

*This paper is based on the presentation "The status of Asia Pacific Reference Frame project" at Eighth Annual Meeting of Asia Oceania Geoscience Society, 8 to 12 August, 2011, Taipei, Taiwan.

purview of the United Nations Regional Cartographic Conference for Asia and the Pacific, and the Reference Frame Sub-Commission 1.3e (SC1.3e) of the International Association of Geodesy (IAG).[6]

Prior to APREF, efforts to densify the ITRF in the Asia-Pacific have included the annual, week-long Asia Pacific Regional Geodetic Project (APRGP) GPS campaigns which have been undertaken throughout the region since 1997 under the umbrella of the PCGIAP.[7] These campaign-style observations have suffered from a myriad of inaccuracies resulting from non-linear coordinate changes associated with the use of tripods,[12] seismic deformation,[4,5] and un-modeled antenna phase variations.[9] These inaccuracies conflict with the requirements of a modern geodetic infrastructure and have motivated the development of APREF as a CORS network rather than continue with the annual epoch-wise ARPGP GPS campaigns.

APREF potentially supports a wide range of scientific applications such as geodynamics research, sea level monitoring and weather prediction. This paper describes the current status of APREF, details the data flow, analysis and combination strategy and provides two example case studies of the applications of APREF products.

2. APREF CORS Network and Data Flow

The APREF project call for participation was released in March 2010,[6,8] and since then GNSS data from more than 400 CORS, contributed by 28 countries, have been incorporated into the network, see Fig. 1. Currently, APREF data are processed by three Local Analysis Centers (LACs) including Geoscience Australia, Curtin University and the Department of Sustainability and Environment, Victoria. The APREF Central Bureau (CB), currently hosted by Geoscience Australia, is responsible for the daily management of APREF. The CB acts as liaison between station operators and analysis centers, providing the necessary station configuration metadata while also ensuring that the data meets the requirements of the analysis. At the CB, the station information, antenna and receiver types in RINEX header are compared to the log files and corrected if necessary in order to guarantee consistency of the station information for further data analysis. However, with the number of stations growing, the need for shorter data latencies and the growing number of applications, the management of the APREF has become a challenge. Although guidelines exist for station equipment, operation and data flow, in practice different institutes use different practices. It is anticipated that as APREF matures data management approaches will increasingly become standardized.

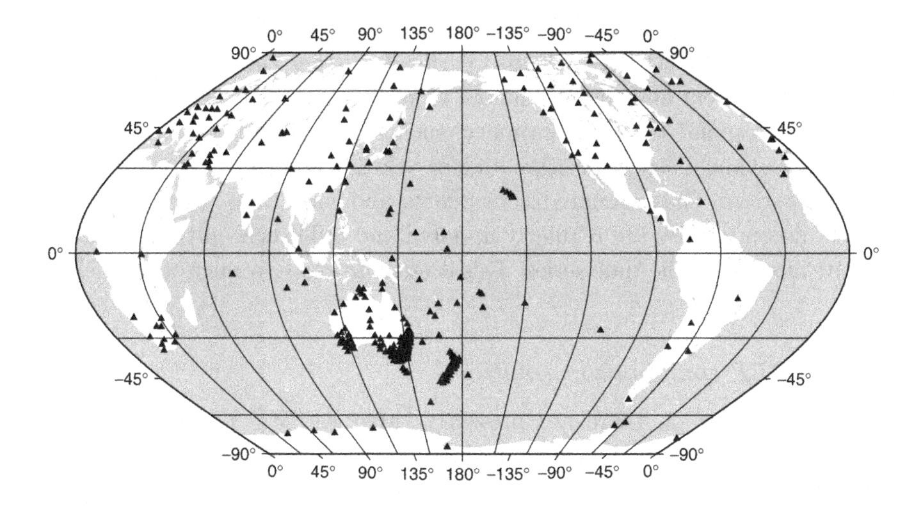

Fig. 1. The APREF CORS Network (as of July 2011). Additional International GNSS Service (IGS) stations, which are used for reference frame alignment, are also shown.

3. Data Analysis and Products

3.1. *APREF CB data analysis*

An important activity of the APREF project is the monitoring and quality control of station coordinates and velocities. To monitor the quality of data as well as the overall performance of stations, routine data analysis is carried out at the APREF CB. This includes the generation of rapid, final and cumulative analysis products. The rapid analysis is performed once a day, based on the RINEX files available, and is generated using the IGS rapid products. The rapid solutions, typically with a latency of one day, provide the necessary information to monitor the performance APREF stations as a tool to assist network management. The daily final solutions are generated using the IGS final products. The latency of the final solutions is approximately two weeks after the release of the IGS final products. The daily final solutions are also combined into a weekly solution that is aligned to the ITRF, specifically IGS08.[1]

A long-term cumulative solution with station coordinates and velocities incorporating the final weekly and the most recent rapid daily solutions is also produced by the APREF CB. This solution is updated on a daily basis depending on the latency of data flow and the IGS rapid product availability. When problems are identified, the APREF CB informs the data provider and/or station operator so that the data quality and stability of stations can be monitored. The APREF CB also performs further time series analysis to identify outliers and discontinuities in station position and velocity. If outliers correspond to a discontinuity, such as seismic displacement and equipment change, a new position and/or velocity is set up and estimated for the station. The estimated coordinates and velocities, together with all the discontinuities are regularly updated and published on the APREF website along with the time series. This is an additional product of APREF.

3.2. *APREF combination solution*

As stated in Sec. 2, there are presently three APREF LACs routinely analyzing data. The LACs generate loosely constrained weekly solutions that are provided to the APREF CB in SINEX format, where the weekly SINEX files are then stacked using the CATREF software[2] to produce the new regularized station coordinates and secular velocities for the global and APREF networks. The combined solutions are aligned to IGS08 using a subset of IGS core sites with a minimum constraint approach.[1]

The weekly SINEX files and updated ITRF coordinate and velocity solutions are published on the APREF website (http://www.ga.gov.au/earth-monitoring/geodesy/asia-pacific-reference-frame.html). However, users should be aware that the published velocities may not be reliable and accurate enough for some applications if less than 2.5 years of data is available for a given station.[3] At this time, a significant portion of the APREF stations have less than 2.5 years of data.

4. Case Studies of Applications of the APREF Project

Two representative examples of applications of the APREF project are given in the following sections.

4.1. *Case study 1: Detection of co-seismic displacement*

GPS data from CORS networks have proven very effective in the accurate measurement of inter-seismic, co-seismic, and post-seismic deformation.[11,13] The two earthquakes which occurred close to Christchurch, New Zealand: Mw 7.1 earthquake 3 September 2010, and Mw 6.3 earthquake 21 February

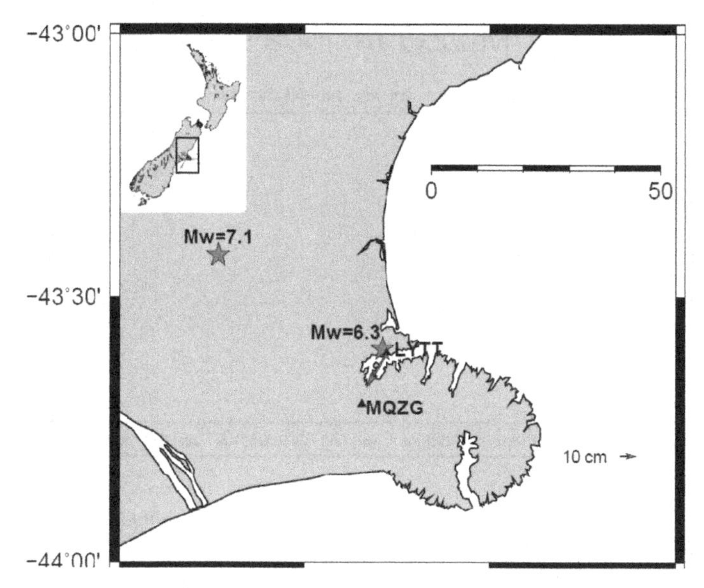

Fig. 2. Locations of earthquake epicentres and the co-seismic displacement of station LYTT caused by the Mw 6.3 earthquake on 21 February 2011 in Christchurch, New Zealand.

2011, provide an example of using APREF analysis to support the measurement of co-seismic displacement. For these earthquakes, based on the solutions of the APREF routine analysis, a 25.1 ± 0.26 cm displacement was detected in SW direction at the CORS of Lyttelton (LYTT), which was 1 km away from the earthquake epicenter of the latter event in Christchurch, as shown in Fig. 2. The displacement was obtained by comparing the station position in daily solutions before and after the earthquake.

From the routine analysis of time series and based on the available data, we also detected the displacements at another APREF GPS station of McQueens Valley (MQZG) in New Zealand, resulting from the two earthquakes, as shown in Fig. 3. The CORS site MQZG lies farther from the epicenter than the station LYTT, which was 12 km away from the epicenter of the Mw 6.3 21 February 2011 earthquake, see Fig. 2. The time series of station MQZG indicates that the station moved approximately 5 ± 0.03 cm north and 10 ± 0.16 cm west as a result of the Mw 7.1 earthquake 3 September 2010; and 5 ± 0.26 cm south and 2 ± 0.16 cm west due to the Mw 6.3 21 February 2011 earthquake. The co-seismic displacement varied in both direction and magnitude for these two earthquakes as well

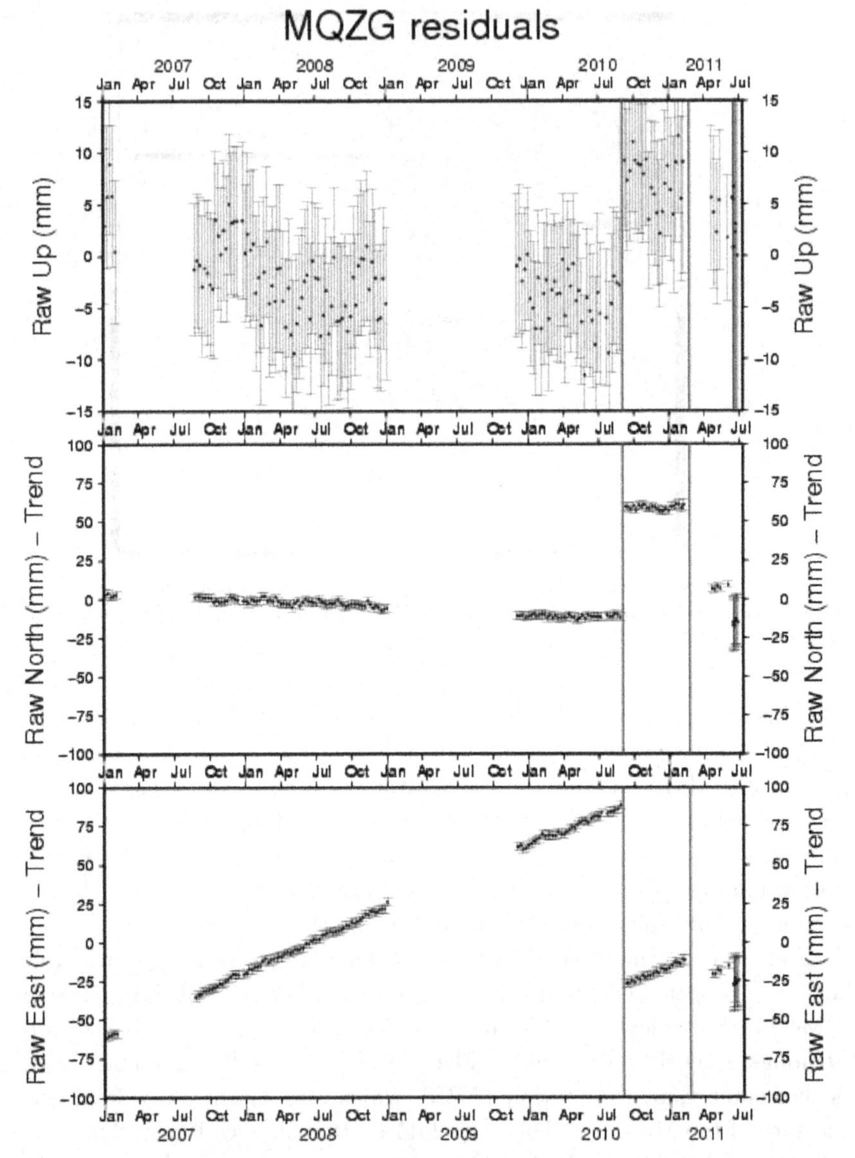

Fig. 3. Time series of GPS station MQZG based on the available data. The abscissa denotes the time in years, and the coordinate are the up, north and east components, respectively. Two vertical lines indicate the timing of the two events, i.e. Mw 7.1 3 September 2010 earthquake and Mw 6.3 21 February 2011 earthquake. The station MQZG moved approximately 5 cm North and 10 cm West resulting from the Mw 7.1 earthquake on 03 September 2010, MQZG lies 46 km South of the epicenter; and moved 5 cm South and 2 cm West caused by the Mw 6.3 21 February 2011 earthquake, MQZG is located 12 km South of the epicenter.

as a significant vertical uplift of 1.5 ± 0.04 cm resulting from the Mw 7.1 3 September 2010 earthquake.

4.2. *Case study 2: Monitoring the stability of CORS site*

Another example of the use of APREF products is stability monitoring of CORS sites by detecting outliers and offsets that degrade the terrestrial reference frame.[10,14] Site stability monitoring is an important part of maintaining a robust regional geodetic framework. With a growing dependency on GNSS data and legal traceability of GNSS, it is important to monitor the stability of a CORS station. Site stability monitoring provides assurance that the CORS data reflects geophysical change in the area, not instability of the monument. By monitoring and characterizing CORS station behavior, users can be assured of high quality data. One representative station is BEE2 in Beenleigh from the APREF network, located in Queensland, Australia. The APREF analysis indicates a vertical offset of 1.5 cm of the station as shown in Fig. 4, which caused by the antenna type changed from TRM33429.00-GP NONE to TRM57971.00 NONE on 17 February 2011. This station event happened but the station operator did not update the log file for another two weeks, highlighting the importance of APREF products as an independent monitoring tool.

5. Future Work

The future APREF work program will include a reprocessing of all available data in IGS08 using IGS reprocessed products so as to improve the reliability and accuracy of APREF station positions and velocities. An APREF cumulative solution will also be made available as the Asia-Pacific contribution to the IAG global dense reference frame solution.

6. Concluding Remarks

In its short history, APREF is already providing critical site specific coordinates and velocities of station in the Asia-Pacific at higher densities than those currently provided by the IGS. Overall, the APREF project offers an improved regional reference frame supporting scientific applications requiring a precise geospatial framework including geospatial information management, sea level rise monitoring, geodynamics research, and disaster

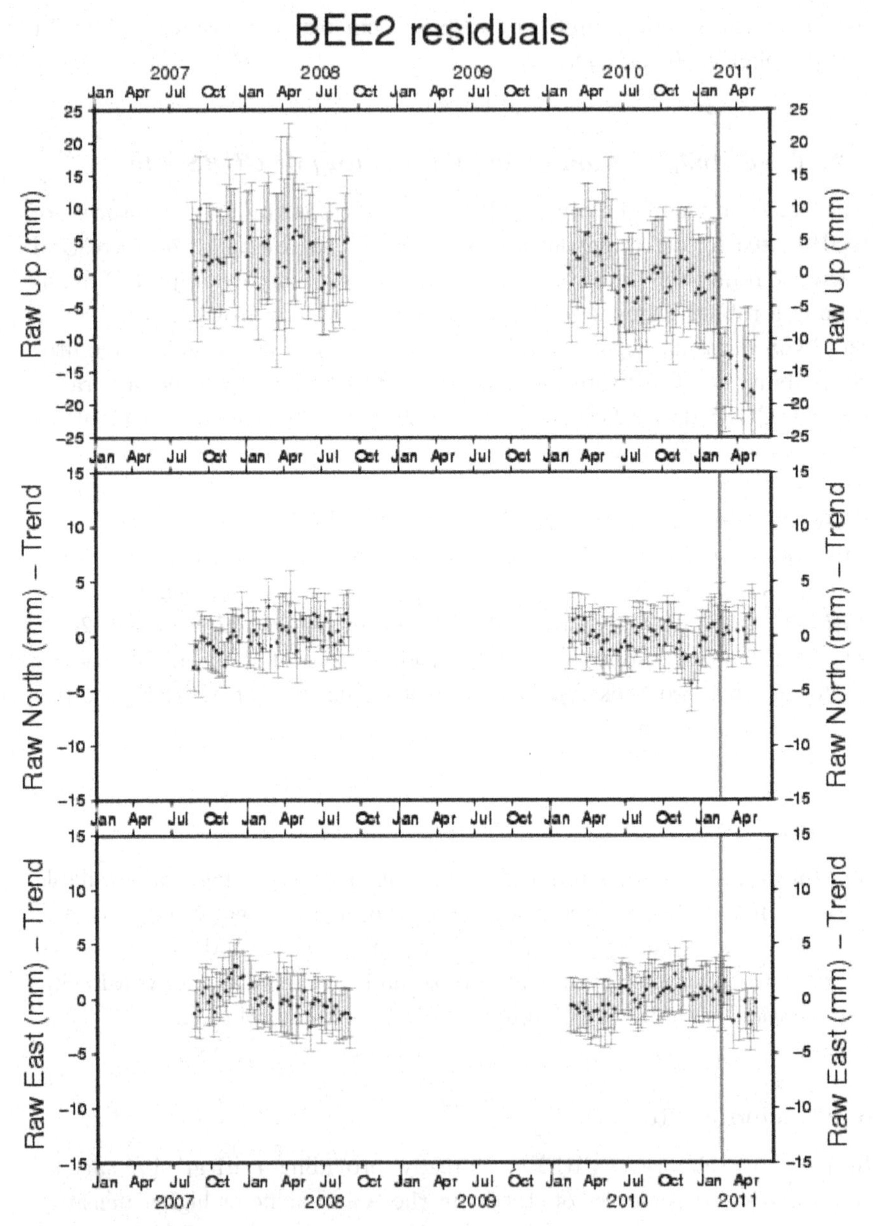

Fig. 4. Time series of station BEE2. The event of antenna type changed at station BEE2 was detected by the APREF routine analysis of the time series. The vertical line indicates the timing of the station event. The abscissa denotes the time in year, and the coordinate are the up, north and east components, respectively. Note that the gap of the time series was caused by the missing data.

prevention and mitigation. APREF is the fundamental basis for the national and regional three-dimensional geodetic infrastructure fully consistent and homogeneous with the ITRF for the Asia-Pacific region.

Acknowledgments

All the participants to the APREF collaborative efforts are gratefully acknowledged, especially the members of the PCGIAP. The CATREF software is kindly provided by Z. Altamimi (IGN, France). The figures were generated by the GMT 3.4 software. The paper is published with the permission of the Chief Executive Officer of Geoscience Australia.

References

1. Z. Altamimi, X. Collilieux and Laurent Métivier, *J. Geodesy*, **85**(8) (2011) 457–473, doi:10.1007/s00190-011-0444-4.
2. Z. Altamimi, P. Sillard and C. Boucher, *J. Geophys. Res.* **107**(B10) (2002) 2214, doi:10.1029/2001JB000561.
3. G. Blewitt and D.Lavallée, *J. Geophys. Res.* **107**(B7) (2002) 2145, doi:10.1029/2001JB000570.
4. Y. Bock, D. C. Agnew, P. Fang, J. F. Genrich, B. H. Hager, T. A. Herring, K. W. Hudnut, R. W. King, S. Larsen, J.-B. Minster, K. Stark, S. Wdowinski and F. K. Wyatt, *Nature* **361** (1993) 337, doi:10.1038/361337a0.
5. J. L. Davis, B. L. Wernicke, S. Bisnath, N. A. Niemi and P. Elo'segui, *Nature* **441** (2006) 1131, doi:10.1038/nature04781.
6. J. Dawson, G. Hu, M. Deo and G. Johnston, The Asia-Pacific Reference Frame (APREF) Initiative. *IAG Commission 1 Symposium 2010*, Marne la Vallee, France, 4–8 October 2010.
7. G. Hu and J. Dawson, The Asia Pacific Regional Geodetic Project (APRGP) GPS Solution (1997–2008). *XXIV FIG International Congress* in Sydney, Australia, 11–16 April 2010.
8. L. Huisman, J. Dawson and P. J. G. Teunissen, *The APREF Project: First Results and Analysis. Coordinates*, 14–18 May 2011.
9. M. A. King, M. Bevis, T. Wilson, B. Johns and F. Blume, *J. Geodesy.* (2011), DOI 10.1007/s00190-011-0491-x.
10. J. Langbein and H. Johnson, *Geophys. Res. Lett.* **22** (1997) 2905–2908.
11. K. P. Larson, Bodin and J. Gomberg, *Science* **300** (2003) 1421–1424.
12. S. Mazzotti, H. Dragert, J. Henton, M. Schmidt, R. Hyndman, T. James, Y. Lu and M. Craymer, *J. Geophys. Res.* **108**(B12) (2003) 2554, doi:10.1029/2003JB002653.
13. T. I. Melbourne, F. H. Webb, J. M. Stock and C. Reigber, *J. Geophys. Res.* **107**(B10) (2002) 2241, doi: 10.1029/2001JB000555.
14. S. D. P. Williams, *J. Geophys. Res.*, **108**(B6) (2003) 2310, doi:10.1029/2002JB002156.

Advances in Geosciences
Vol. 31: Solid Earth Science (2011)
Eds. Ching-Hua Lo *et al.*

TOWARDS SYNERGY OF VLBI AND GNSS GEODETIC TECHNIQUES IN GEOLOGICALLY DYNAMIC NEW ZEALAND

H. TAKIGUCHI*, S. GULYAEV, T. NATUSCH,
S. WESTON and L. WOODBURN
Institute for Radio Astronomy and Space Research,
Auckland University of Technology,
Duthie Whyte Building, Ground Floor, 120 Mayoral Drive,
Auckland 1010, New Zealand
**hiroshi.takiguchi@aut.ac.nz*
www.aut.ac.nz

Auckland University of Technology (AUT) has built the first research capable radio telescope in New Zealand designed for geodetic VLBI. Here, we discuss possible inter-technique (VLBI and GNSS) synergies capable of increasing the reliability of the national geodetic infrastructure. While a GNSS station network can generate a relative local velocity model, VLBI is able to directly link this local model to the ITRF. Development of synergy between these two geodetic techniques is, therefore, one of the central tasks in sustaining the New Zealand Geodetic Datum 2000 and maintaining its link to the ITRF. In addition, we briefly compare New Zealand and Japanese synergetic approaches and discuss differences and commonalities.

1. Introduction

Auckland University of Technology (AUT) has constructed a 12-m geodetic radio telescope in New Zealand near Warkworth, 60 km north of Auckland (Fig. 1). It was launched on 8 October 2008 as New Zealand's first research capable radio telescope. Recently, it became a network station of the International Very Long Baseline Interferometry (VLBI) Service for Geodesy and Astrometry (IVS). The radio telescope is colocated with a Global Navigation Satellite System (GNSS) station operated by Land Information New Zealand (LINZ). Through participation in the observational programs of the IVS, New Zealand will be able to contribute towards establishment of the global reference frame.

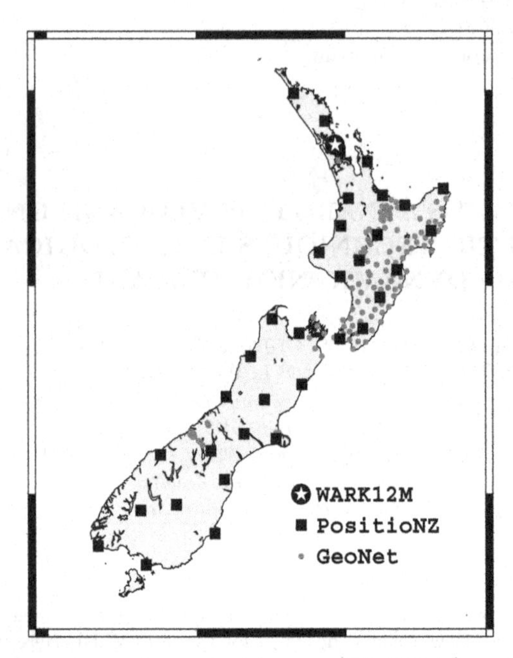

Fig. 1. The location of the 12-m radio telescope (WARK12M) is shown with a star. Stations of PositioNZ and GeoNet (except those in the Chatham Islands and Antarctica) are shown with squares and dots.

In particular, this new telescope contributes significantly to expansion of the IVS network of radio telescopes in the southern hemisphere, where it is notably sparse.[2] The New Zealand Geodetic Datum 2000 (NZGD2000)[6] is defined by its relationship to a dynamic global reference frame at a specified epoch (2000.0) and does not include coordinate changes that occur (in geologically dynamic New Zealand) as a result of earthquakes and tectonic motions. It was established with the use of GNSS data alone in mind, because GNSS was the only space geodetic technique in New Zealand before launch of the geodetic radio telescope. By combining the data from GNSS and VLBI, which is capable of linking the International Celestial Reference Frame (ICRF) and the International Terrestrial Reference Frame (ITRF), New Zealand will be able to contribute to monitoring of the difference between the NZGD and the ITRF.

In this report, we describe the synergistic relationship that we aim at and show the first synergistic result from combined VLBI and GNSS data. In addition, we discuss the international experiment using this telescope currently in progress.

2. Specification of 12-m Radio Telescope at Warkworth

Figure 2 shows the 12-m radio telescope at Warkworth (hereafter WARK12M). It was manufactured by Patriot Antenna Systems (now Cobham Antenna Systems), USA. The antenna specifications are provided in Table 1. The radio telescope is designed to operate at S and X band and is equipped with an S/X dual-band dual-polarization feed. From inception, it was designed for the purpose of doing geodetic VLBI. It is also considered as a prototype of a New Zealand station of the Square Kilometre Array (SKA) project that Australian and New Zealand jointly bid for.

Fig. 2. Warkworth 12-m radio telescope (WARK12M).

Table 1. Specifications of the Warkworth 12-m radio telescope (WARK12M).

Antenna type	Dual-shaped Cassegrain
Manufacturer	Cobham/Patriot, USA
Main dish Diam.	12.1 m
Secondary refl. Diam.	1.8 m
Focal length	4.538 m
Surface accuracy	0.35 mm
Pointing accuracy	18"
Frequency range	1.4–43 GHz
Mount	alt-azimuth
Azimuth axis range	90° ± 270°
Elevation axis range	6° to 88°
Azimuth axis max speed	5°/s
Elevation axis max speed	1°/s
Main dish F/D ratio	0.375

The antenna is equipped with a digital base band converter (DBBC) developed by the Italian Institute of Radio Astronomy, a Symmetricom Active Hydrogen Maser MHM-2010 (75001-114), and Mark5B+ and Mark5C data recorders developed at MIT Haystack Observatory.

The support foundation for the antenna is a reinforced concrete pad that is 1.22 m thick by $6.7 \times 6.7\,\mathrm{m}^2$. The ground that the foundation is laid on consists of weathered sandstone/mudstone, i.e. is of sedimentary origin, laid down in the Miocene period some 20 million years ago. The pedestal is essentially a steel cylinder of 2.5 m diameter. It supports the antenna elevation axis which is at a height of approximately 7.1 m above ground level. Apart from the pedestal, all other components of the antenna (the reflector and feed support structure) are constructed of aluminium. The radio telescope is directly connected to the regional broadband network Kiwi Advanced Research and Education Network (KAREN).[4,13] KAREN in turn is connected to Australia and USA by the Southern Cross cable. Currently, this network connectivity is 1 Gbps.

3. Synergy between VLBI and GNSS in New Zealand and Japan

When considering the synergy of VLBI and GNSS, it is useful to have an example to follow. We believe that Japan is a good model for New Zealand as the geological situation in these two countries is very similar: Both are located on plate boundaries, both countries are seismically active, both are subject to crustal deformation and volcanic activity[1,8,12] (Fig. 3). Also, the two countries have VLBI stations, GNSS networks and similar geodetic datums.

New Zealand has two continuous GNSS networks (Fig. 1). One is called PositioNZ and it is maintained by LINZ. PositioNZ consists of 31 stations located across mainland New Zealand, plus two stations in the Chatham Islands (roughly 800 km east of Christchurch) and three in the Ross Sea region of Antarctica. The PositioNZ network was used to construct New Zealand's geodetic system (NZGD2000) and is used for surveying and mapping. The other GNSS network is called GeoNet. GeoNet is a system designed for monitoring earthquakes, volcanic unrest, land deformation, geothermal activity and tsunami. Therefore, each GeoNet station consists of not only GNSS itself, but also a seismometer, accelerometer, and where appropriate tide gauge, and a sea level pressure measurement device. The GeoNet GNSS stations are mainly located in active volcanic areas of New Zealand's North Island.

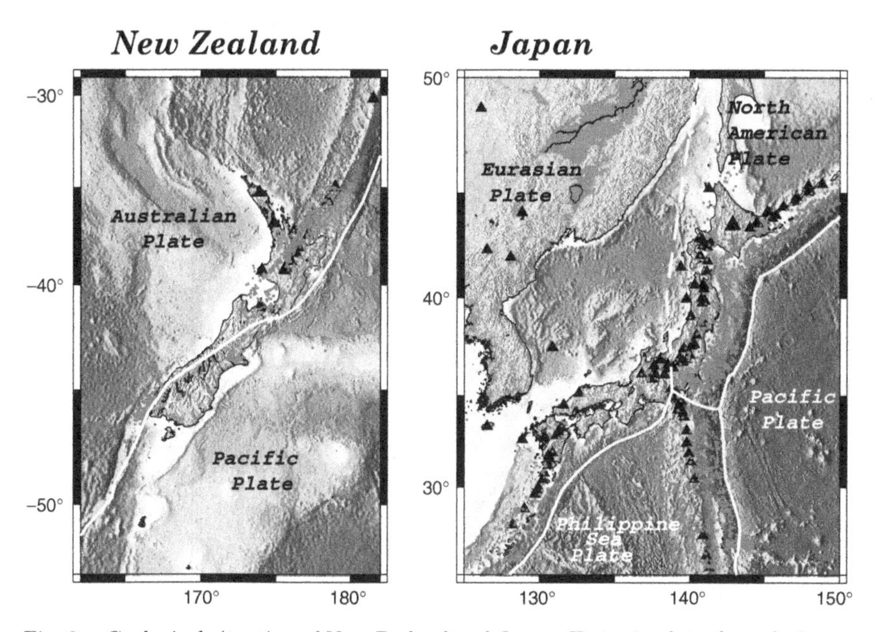

Fig. 3. Geological situation of New Zealand and Japan. Tectonic plates boundaries are shown with yellow curves. Red dots indicate the epicenter ($M \geq 1$, from January 2008 to December 2010) according to the USGS/NEIC earthquake catalog. Black triangles indicate the location of the major volcanoes.

The GNSS network in Japan is also called GEONET, which stands for the GPS Earth Observation Network System (Fig. 4). This network is maintained by the Geospatial Information Authority of Japan (GSI) and consists of over 1,200 GNSS stations. While the number of stations in Japan and New Zealand is different, each has a continuous GNSS network spread across the whole countries territory.

Geodetic datums adopted in New Zealand and Japan are slightly different. While New Zealand has a semi-dynamic system (NZGD2000), Japan has adopted a static system (JGD2000).[3,5] Recently, Japan introduced a semi-dynamic correction in reference to New Zealand's semi-dynamic system.[10] The decisive difference between the two geodetic systems is, however, that the New Zealand geodetic datum was defined solely by GNSS data, whereas the Japanese geodetic datum was defined primarily by VLBI and supplemented by GNSS.

The important synergetic relationship between VLBI and GNSS has been clearly established by the Japanese experience and provides an excellent model for future progress in New Zealand. Figure 5 is a summary of the synergetic relationship between VLBI and GNSS in Japan modified after

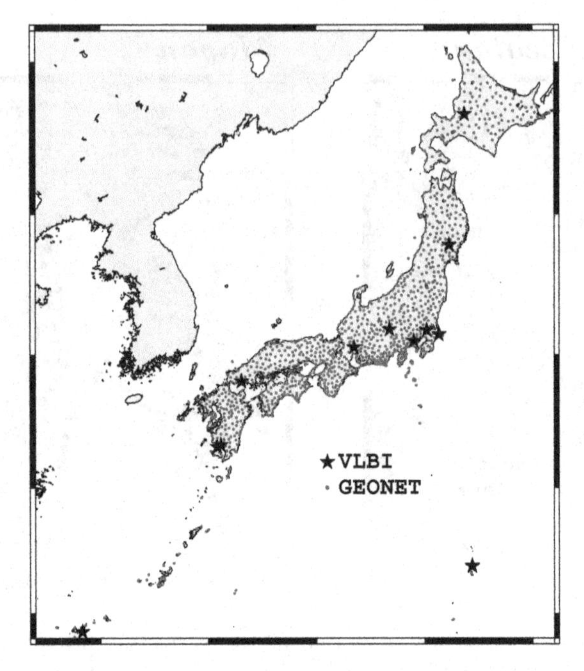

Fig. 4. The GEONET network and VLBI stations in Japan.

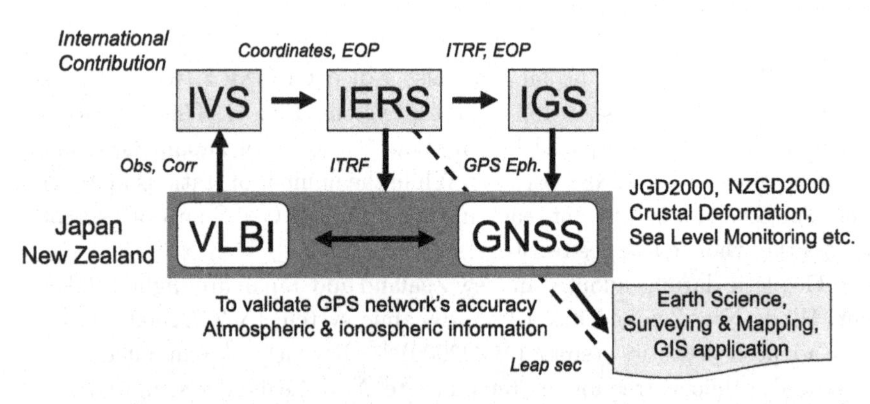

Fig. 5. Synergetic relationship between VLBI and GNSS. Modified after Tsuji *et al.*[9]

Tsuji *et al.*[9] VLBI contributes to determination of Coordinates and Earth Orientation Parameter (EOP) solution of the IVS. The International Earth Rotation and Reference Systems Service (IERS) combines Coordinates and EOP solutions from each space geodetic techniques; VLBI, GNSS, and SLR. The ITRF and EOP solutions from IERS are very important for the precise determination of GNSS satellite orbits by the International GNSS Service (IGS). In Japan, IGS products are used for analyzing daily GNSS data to monitor crustal deformations. The GNSS network is also used for surveying, mapping, and GIS application. VLBI contributes towards validation of the GNSS network accuracy. GNSS, in turn, provides atmospheric and ionospheric information for VLBI. We argue that establishment of similar synergetic relationships should be seriously considered and adopted in New Zealand.

4. First Synergistic Result

To determine the initial coordinate of our VLBI reference point, we conducted a GNSS survey in collaboration with New Zealand Institute of Geological and Nuclear Sciences (GNS Science) and LINZ in March 2010. A real-time kinematic (RTK) GPS method was used to derive the position with respect to the co-located GNSS station WARK (Fig. 6).

WARK was established in November 2008 at the radio telescope site and is one of the PositioNZ network stations. The RTK reference receiver was set up in an arbitrary location with clear sky view and was configured to record raw observations in addition to transmitting real-time corrections.

Fig. 6. The GNSS station WARK (in front of the picture) and the VLBI network station WARK12M.

The RTK station was later post-processed with respect to WARK, and all RTK rover-surveyed positions were subsequently adjusted relative to the updated reference position. The details of this survey were described in Gulyaev *et al.*[4] and Petrov *et al.*[7] As a result, GNSS coordinates of the VLBI reference point were obtained (Table 2).

After the coordinates of the VLBI reference point were determined, the VLBI technique was used to estimate the coordinates of the Warkworth radio telescope. As shown in Fig. 7, seven stations (WARK12M and ASKAP-29, Ceduna, ATCAPN5, Mopra, Parkes, Hobart26 that belong to the Australian Long Baseline Array (LBA)) participated in a 7 h long VLBI experiment on 07 May 2010. Data was processed by L. Petrov (Table 2). These observations were conducted in L-band in conjunction with the first astronomical observations of the first ASKAP antenna.[11] Because we could not reduce the effect of ionosphere path delay, the systematic error was

Table 2. Coordinates of the Warkworth 12-m antenna.

Study	X	Y	Z
RTK GPS	−5115324.5±0.1	477843.3±0.1	−3767193.0±0.1
Petrov *et al.* (2011)	−5115325.55±0.10	477842.95±0.05	−3767194.41±0.09
IVS, by O.Titov	−5115324.41±0.02	477843.31±0.02	−3767192.93±0.06

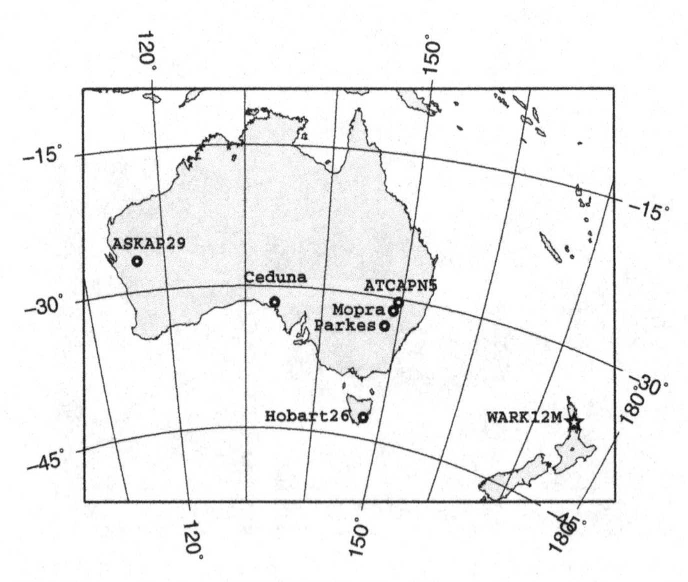

Fig. 7. The LBA network and WARK12M. The LBA stations are shown with circles, and the WARK12M radio telescope is shown with a star.

expected to be high. The details of this experiment were described in Petrov *et al.*[7]

WARK12M has now been participating regularly in IVS sessions from the beginning of 2011. We also obtained an estimate of the coordinates of WARK12M from analysis of IVS R1 sessions. This result agrees well with GNSS coordinates with less than 10 cm difference for each component (Table 2). These latter coordinates are currently used as the coordinates of WARK12M. This series of coordinate measurements can be considered as the first synergetic result between VLBI and GNSS in New Zealand.

5. Summary

We summarized the aim of the synergy of VLBI and GNSS in New Zealand and presented the first synergistic result; determination of the position of the Warkworth 12-m radio telescope by using GNSS and geodetic VLBI techniques. Establishment of VLBI and synergistic relationships with GNSS can be achieved with the help of the model adopted in Japan. As our first step towards this goal, we are working on the ultra-rapid EOP experiment in cooperation with Japanese institutes, GSI and NICT. The first fringes between Warkworth 12-m and Tsukuba 32-m antennas have been obtained in both S band and X band (Fig. 8).

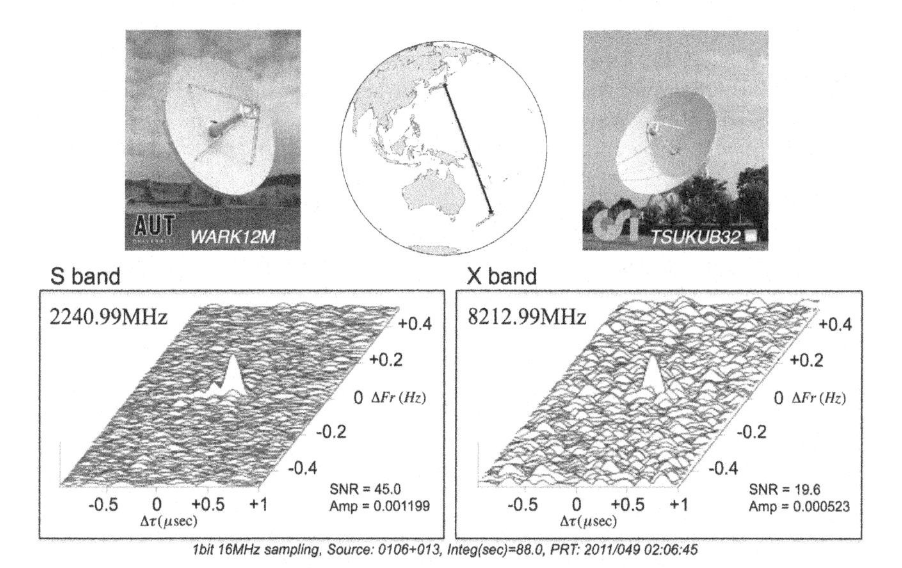

Fig. 8. The Ultra-rapid EOP measurement for the WARK12M (left) and TSUKUB32 (right) baseline. The first fringes of this baseline for both S and X band are presented.

Acknowledgments

The authors would like to acknowledge Leonid Petrov, Neville Palmer, Dave Collett and Oleg Titov, and organisations and institutions such as IVS, IGS, LINZ, GNS Science, the Commonwealth Scientific and Industrial Research Organisation (CSIRO), GSI, NICT and KAREN.

References

1. K. R. Berryman, Y. Ota and A. G. Hull, *Quaternary Int.* **15–16** (1992) 31–45.
2. B. Blick and S. Gulyaev, *Surv. Q.* **58** (2009) 23–25.
3. Geographical Survey Institute, *Bull. Geogr. Surv. Inst.*, **50** (2004) 33–36.
4. S. Gulyaev, T. Natusch, S. Weston, N. Palmer and D. Collett, *IVS 2010 Annual Report*, 2010.
5. S. Matsumura, M. Murakami and T. Imakiire, *Bull. Geogr. Surv. Inst.* **51** (2004) 1–9.
6. M. Pearse, *OSG Technical Report*, **5** 2000.
7. L. Petrov, C. Phillips, T. Tzioumis, B. Stansby, C. Reynolds, H. Bignall, S. Gulyaev and T. Natusch, *Publ. Astron. Soc. Aust.* **28**(2) (2011).
8. Rafferty, P. John, *Plate Tectonics, Volcanoes, and Earthquakes*, (Rosen Educational Publishing, 2010).
9. H. Tsuji, T. Tanabe, H. Kawawa, K. Miyagawa, K. Takashima, S. Kurihara, Y. Fukuzaki and S. Matsuzaka, *J. Geodesy*, Proceedings of the IUGG General Meeting, 2003.
10. H. Tsuji and K. Komaki, *Bull. Geogr. Surv. Inst.* **52** (2005) 1–11 .
11. A. K. Tzioumis, S. J. Tingay, B. Stansby, J. E. Reynolds, C. J. Phillips, S. W. Amy, P. G. Edwards, M. A. Bowen, M. R. Leach, M. J. Kesteven, Y. Chung, J. Stevens, A. R. Forsyth, S. Gulyaev, T. Natusch, J.-P. Macquart, C. Reynolds, R. B. Wayth, H. E. Bignall, A. Hotan, C. Hotan, L. Godfrey, S. Ellingsen, J. Dickey, J. Blanchard and J. E. J. Lovell, *Astron. J.*, **140** (2010) 1506–1510.
12. L. M. Wallace, J. Beavan, R. McCaffrey and D. Darby, *J. Geophys. Res.* **109** (2004) B12406, doi:10.1029/2004JB003241.
13. S. Weston, T. Natusch and S. Gulyaev, *Proceedings of the 17th Electronics New Zealand Conference*, 2010.

Advances in Geosciences
Vol. 31: Solid Earth Science (2011)
Eds. Ching-Hua Lo *et al.*

REINTERPRETATION OF HISTORICAL EARTHQUAKES DURING 1929 TO 1931, MYANMAR

HLA HLA AUNG
Myanmar Earthquake Committee
MES Building, Hlaing University Campus, Yangon, Myanmar
hhlaaung@gmail.com

The years between 1929 to 1931 are significant years for Myanmar region, with a series of earthquakes starting with Swa Earthquake in 1929 to Phyu Earthquake in 1931. Major earthquakes that occurred during the years 1929 to 1931 in the past are in a remarkably linear distribution along the Sagaing fault. The area along the Sagaing fault falls in Zone 1 of the Seismic Zones of Myanmar. A detailed study of the Sagaing fault through the satellite images and aerial photographs reveals a zigzag nature consisting of alternate long NNW–SSE striking and short ENE–WSW striking fracture lineaments. The former occurs in a right-handed en-echelon strike-slip pattern which are cut by the latter being expressed as a normal fault character. The presence of dominant ENE-WSW striking normal faults suggest that the area has been experiencing NNW-oriented regional tensional stress. Locations of historical earthquakes for the years of 1929 to 1931 have defined two ruptured segments viz. Segment (1) and Segment (3). Since 1929, earthquakes were mainly occurring in the southern and northern parts of the Sagaing fault. Therefore, it can be inferred that until end of 1991, the seismic activity was mostly confined on Segment (1) and Segment (3). To better examine the relationship between the seismicity and tectonics, major historical earthquakes of Pyinmana (1931), Pegu (Bago) (1930), Phyu (1930) and Swa (1929) are grouped as Segment (1) and Tagaung (1991), Mandalay (1956), Wuntho (1946) and Kamaing (1931) as Segment (3). The remaining unruptured fault length about 200 km in between these two segments is Segment (2). A detailed study of characteristics of each earthquake show that periodic movement along the normal lineaments has given rise to seismicity, and the location of earthquake events coincides with zones of lateral dislocation of Sagaing fault and occurrence of localized ridges or localized basins in the fault zone.

1. Tectonic Setting

Myanmar is situated in the South East Asia (SEA) region, which is located on the boundary of the Indo-Australia and Eurasia plates. Myanmar is composed of two different evolving continents of the Burma plate and the Indochina plate. The two continents collided and were welded by the magma during the late Cretaceous-early Eocene. The Sagaing fault is a major continental strike-slip fault and a continental transform plate boundary between the Burma plate and the Indo-China plate.[8,9,18] Continental transforms are complicated and are active seismic zones. Historical earthquakes that occurred along the Sagaing fault and focal mechanism solution of recent seismicity show the complexity of faulting. This has included strike-slip faulting on the Sagaing fault itself, thrust faulting for events like the Pagan Earthquake (December, 1975) and strike-slip faulting like the Taungdwingyi Earthquake (September, 2003) due to the regional extension in Central Myanmar Basin. The axes of anticlines and synclines in the Central Myanmar Basin display acute angles of about 40° to the strike of the Sagaing fault in the south and fold axes swing clockwise to the north–northwest direction in the north.[2] The Sagaing fault zone is a large dextral shear band extending at least 100 km on both sides of the main fault.[22] The breadth of the continental plate boundary zone has important implications for seismic hazard within them. The Central Myanmar Basin has classic continental rift basin structures which are developed as a result of oblique subduction of the India plate beneath the Burma plate and are arranged in basin-and-uplift configuration. The rift basins in the Central Myanmar Basin, existence of normal faults and strike-slip faults which bound these basins, and distribution of seismicity show the presence of an extensional broad zone about 250 km wide in Myanmar. This en-echelon basin system extends more than 2000 km from the East Himalayan Syntaxis to the north, to the Andaman Sea to the south. Pivnik *et al.*[24] report that this extensional broad zone was subject to NNW-directed extensional deformation in the Miocene, followed by ENE-directed Pliocene–Pleistocene transpressional deformation. Incipient motion along the Sagaing fault was estimated to be between 4 and 5 Ma, at the most. This fault absorbs only part of the dextral India/Sundaland motion, which could be estimated between 18 and 25 mm/yr.[30] These estimations imply a total cumulated dextral offset bracketed between 100 km and 150 km, at the most.[22] Historical earthquakes on the Sagaing fault and recent seismicity within the fault zone show that although most of the motion occurs along the Sagaing fault, or nearby parallel or subparallel faults, some portion of the motion occurs elsewhere within the fault zone.

2. Introduction

This paper is based on the historical accounts of earthquakes from the "Geology of Burma" by Chhibber, 1934 and "Major Earthquakes of India" by Satyabala, 1998; and other related literature. Although Brown *et al.*,[5] Chhibber,[6] and Satyabala[28] suggested that the historical earthquakes for the years 1929 to 1931 were caused by the movement of the Sagaing fault, no geological evidence on the relation between surface topography and seismic amplification for these earthquake events have been presented until this paper. Reinterpretation of historical earthquakes for the years 1929 to 1931 have been carried out by the author for the first time in Myanmar. This reinterpretation of historical earthquakes along the Sagaing fault is only a beginning. Global Positioning System (GPS) data and paleoseismotectonic studies are greatly needed to conduct on the Sagaing fault in a systematic manner. Dey,[14] Win Swe[38] (1972), Le Dain *et al.*,[21] and Rangin *et al.*[26] have mapped the entire fault trace based on studies of land sat images. The Sagaing fault plays a major role in the tectonics of Myanmar and several $M > 7$ earthquakes occurred along the fault in the last century. To correlate the tectonics and the seismicity, the Sagaing fault has been mapped using 1:24,000 scale aerial photographs, 1:63,360 (1 inch = 1 mile) scale topographic maps and satellite images. Many tectonic geomorphic features have been observed along the Sagaing fault. All these observations and other data from published literature testify to the fact that the Sagaing fault is a major continental strike-slip fault and a continental transform fault.

3. Tectonic Signature of the Sagaing Fault

The Sagaing fault is a major strike-slip fault, which has long and straight traces across the entire length of Myanmar for 1,000 km. The Sagaing fault is linked with the Central Andaman spreading center to the south (Fig. 1).[8] The fault is distinct on satellite images but is much more difficult to observe between Mandalay and Naypyidaw. To the north, the fault branches in various splays north of Thabeikyin and terminates in the Jade Mine area into a compressive horse tail structure 200 km wide from east to west. The plate boundary has moved from east to west due to the westward push of crustal material from the east.[22] In the central part, the trace of the fault is rather linear and simple and composed of several fault segments arranged in right-handed, en echelon pattern. In the south, in the Mottama Gulf, the fault terminates as horse tail extensional system with two main

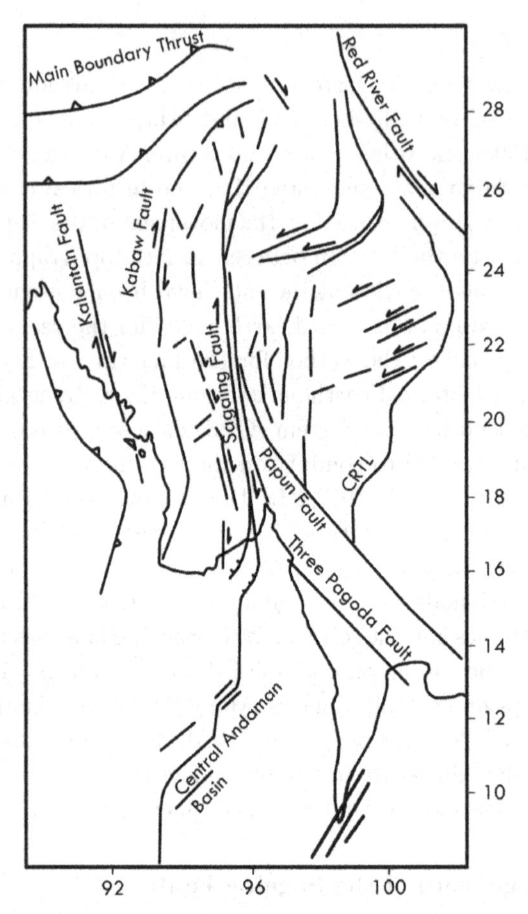

Fig. 1. Map showing tectonic lineaments in Myanmar from satellite images. The Sagaing fault is a major strike-slip fault which have long and straight traces across the entire length of Myanmar for 1,000 km. The Sagaing fault is linked with Central Andaman spreading center to the south.

fault branches a few kilometers apart. The western branch of the fault, West Sagaing Fault terminates southwards in a cluster of normal faults separated by a series of pull-apart basins and a system of oblique transform faults that connect southwards with the Andaman spreading center.

Within the neotectonic framework of Myanmar, the Sagaing fault is relatively more westerly (NNW–SSE) in the south than N–S linear in its central part between Pyinmana and Thabeikyin, a distance estimated to be 300 km. It then swings to the easterly direction (NNE–SSW) again further in the north of Thabeikyin. When the Sagaing fault trends differ from the plate motion, dip slip faulting occurs. Most of the NNW-striking faults that dominate in the area appear to be presently active, as shown by the recent seismicity which includes historical earthquakes. A detailed study along the fault using the satellite images and aerial photographs reveals a zigzag nature consisting of alternate long NNW–SSE direction and short ENE–WSW trending lineaments. The former occurs in a right lateral en-echelon pattern being connected by latter. The longer faults are strike-slip faults and the latter are normal faults. From the studies of fault characteristics along the Sagaing fault, the faults record both strike-slip deformation and dip-slip displacements that appear to reflect two episodes of strike-slip deformations. The first involved a combination of strike-slip motion and extension on north–northwest trending faults, leading to the formation of localized pull-apart basins. The second involved strike-slip motion and compression forming localized folds and thrusts, possibly as a result of a change in the direction of motion in the transformation.

In fault zones where strike-slip faults are continuous, the strike of the faults may locally form a simple linear trend following a small circle on the Earth's surface. In these areas, the curvature of the fault plane creates zones of localized shortening and extension according to whether the two sides of the bend coverage or diverge.[8,17] These zones are similar to those that form in step-overs — pull-apart basins, zones of subsidence and deposition, and normal faults characterize releasing bends, whereas restraining bends display thrust faults, folds, and push-ups. These ranges have been localized and uplifted in response to a combination of dextral motion and compression across a portion of the fault that strikes more westerly than the general strike of the fault system. At converging bend, localized zone of subsidence occur due to the combination of extension and dextral motion. Extensional deformation has been followed by compression in late Miocene to Pliocene–Pleistocene. Rhomb shaped grabens, elongate pull-apart basins, and steep normal faults have formed where the fault segments step to the right forming Kabauk In, Zwedeik In, Pagan In, Shwedan In in the southern part of the Sagaing fault, whereas Yega In and Indawgyi Lake in the northern part were formed as well as many others in different parts of the fault zone.

4. The 1929–1931 Earthquake Events and their Site-Specific Characteristics

Coseismic deformations were summarized from the publications of "Geology of Burma" by Chhibber, 1934 and "The Major Earthquakes of India" by

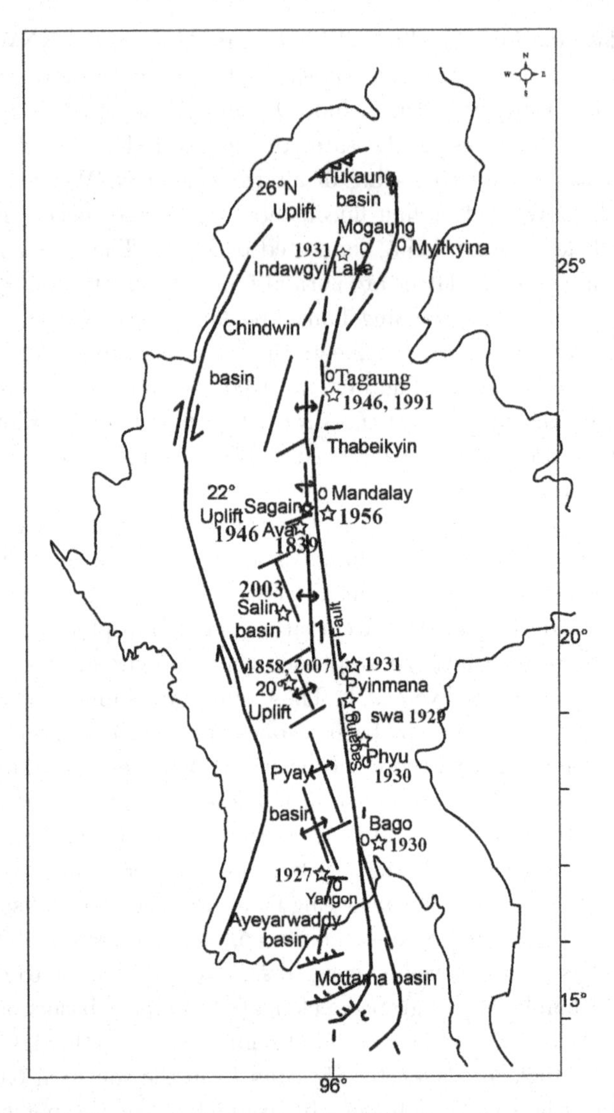

Fig. 2. Map displaying intraplate seismicity in rift basins and occurrences of historical earthquakes along the Sagaing fault in Myanmar (Data from Chhibber,[7] Satyabala,[28] and USGS earthquake catalog).

Satyabala, 1998. The years 1929 to 1931 are significant years in the region of Myanmar starting with an earthquake on 8 August 1929 in Swa to Phyu on 23 September 1931. These earthquakes occurred in a remarkably linear distribution most of which were along the Sagaing fault (Fig. 2). From the studies of aerial photographs, satellite images and digital elevation model of location of each event, local site-specific characteristics of the area where earthquake occurred with large magnitude are quite significant for the amplification of seismic waves. Studies of site-characteristics of each event and co-seismic surface deformation indicate that topography of epicentral area is the key for seismic amplification.

4.1. The Swa earthquake of 8 August 1929 was the first of a series of earthquakes within the years 1929 to 1931 to affect the region of Myanmar spread linearly over about 250 miles (~400 km). The earthquake occurred with its epicenter about six miles to the northwest of Swa, a small town situated at (19°13′N, 96°14′E) in the northern part of the Toungoo district. At Pyinmana, about 30 miles north, detail of damage was not available to assign an intensity but at Yamethin, 83 miles to the north of Swa, the shock was still strong enough to sway buildings and to displace hanging objects. In the setallite images, a long linear fault scarp has been observed in the west of Swa.

4.2. Pegu (Bago) earthquake occurred on 5 May 1930 that practically destroyed the town of Pegu causing great loss of life. It was felt with continually diminishing intensity, over a land area of approximately 220,000 square miles (570,000 sq. km.). The epicenter of this earthquake was at (17°N, 96.5°E) and Ms = 7.3[1] and origin time 13:45:27 hours GMT.[16] It is in the Bago deltaic alluvial plain, in the south of the Bago–Sittoung canal. In Rangoon, most of the damage was done to the southern portion of the city that was built on alluvium and "made-ground" and the damage was less further north where the ground was composed of Tertiary rocks.

The most common liquefaction features are many large cracks and block fissures that appeared in many areas such as Thanatpin in north of Bago and Kyauktan, Thongwa and Thanlyin in south of Bago along north–south direction. Sand blow and sand vents which erupted water and sand were found in many places and even reached the coast. Rising of local ground level and damming of the stream are observed in the Bago area.[28] Epicentral tract of the Bago earthquake lies in the Bago plain mainly made up of deltaic alluvium, in the south of the Bago–Sittoung canal and it is the sites of occurrence of localized pull-apart basins such as Kabauk In, Zwedeik In, Pagan In and Shwedan In. Epicentral location is the site of occurrence of Kabouk In (In is local name for a lake). It is a pull-apart basin marked by irregular depressions and oblique normal faults and formed in an extensional step-over between the fault segments. A large earthquake tends to nucleate from the tip of the segment and then propagate along the fault plane where the accumulated stress is high to incur slip on the fault plane. Seismicity is consistent with the transfer of slip on one fault onto another fault segment to the right in this area (Fig. 3)

4.3. Phyu earthquake at 1:22 a.m. on 4 December 1930 hit the town of Phyu in Myanmar and was shaken by a violent earthquake Field investigations indicated an epicenter of this shock at between four and six miles to the WSW of Phyu, but in the town it was strong enough to cause major destruction. This earthquake was felt throughout a great part of Myanmar, southern and northern Shan State and Thailand over an area of 220,000-sq. mi. The epicenter of this earthquake was at (18°N, 96 1/2°E), origin time 18:51:44 GMT of 3 December 1930 and Ms = 7.5.[1] The most affected area is in a small limited area to the west and WSW of Pyu (18°24′N, 96°24′E), in the vicinity of Kindangyi (18°28′N, 96°24′E) where the earthquake was extremely severe. The peak Khengdan may be taken as approximately the center of the most shaken area. The approximate limits of the epicentral tract to the north and south of Kindangyi respectively are regions in the west of Zeyawadi (18°33′N, 96°26′E) and northwest of Penwegon (18°12′N, 96°36′E); they give some indication of the approximate limits of the epicentral tract. Mud and water had spouted to heights of 18–20 feet, like a fountain, from the cracks. Fissuring of alluvial ground and issue of sand and water from craterlets are prominent features in all areas felt by this severe earthquake. At Phyu area, a diverging bend from the main Sagaing fault is observed on the satellite images. When there is diverging bend in a strike-slip fault zone, bordering country rocks are refined to adjust stress buildups by shortening. These are called restraining bends and are common places to respond to crustal uplift of thickened blocks formed by folding or thrusting (Fig. 4).

4.4. The Kamaing earthquake occurred on 28 January 1931, with a magnitude Ms = 7.6.[16] It was a disatrous earthquake which caused developement of big fissure sometimes several feet long in the alluvial area, accompanied with spoutings of sand and water. Epicenter was in the east of the Indawgyi Lake, very close to the Sagaing fault. The epicentral tract is very hilly and its everage height is between 2000 and 3000 feet above sea-level. Its maximum height is 4,982 feet above sea-level. The hill slopes were scarred by big cracks and block fissures were also observed. Big fissures with several hundreds feet long developed in the alluvial tracts, accompanied by spouting of sand and water. Numerous landslips and rockfalls occurred in the epicentral tract.

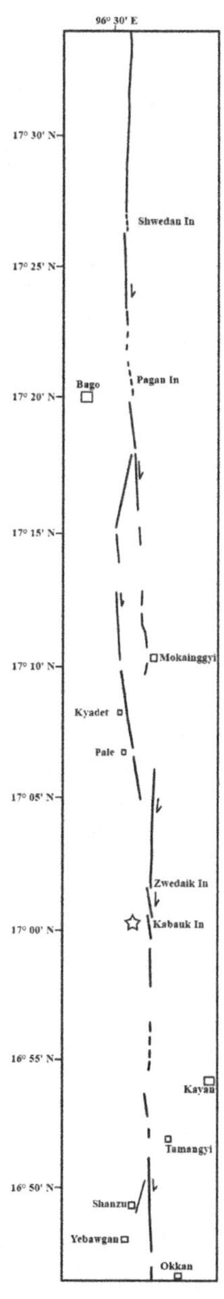

Fig. 3. Detailed fault map of the study area between latitude 16°45′N and 17°35′N, interpreted from 1:24,000 scale aerial photographs and 1:63360 scale topographic maps showing several pull-apart basins (local name = In): Kabauk In, Zwedaik In, Pagan In, Shwedan In, showing coincidence of the epicentral location of Bago earthquake and occurrence of Kabauk In (a localized pull-apart basin) in lower part of the fault map.[33]

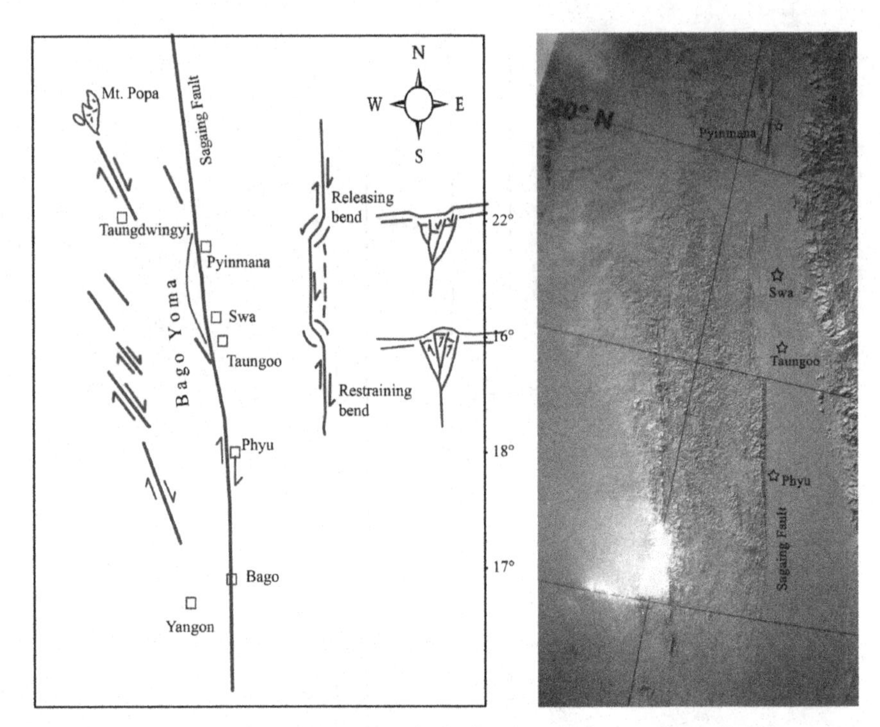

Fig. 4. Releasing and restraining bends in Sagaing Fault zone near Pyinmana area and Taungoo–Phyu area. Releasing/restraining bends are observed on satellite images along the Sagaing fault between latitude 18° N and 22° N. Epicentral location of historical earthquakes exceeding magnitude 7 (Swa, 1929, Phyu, 1930, Pyinmana, 1931) are shown on the map.

The Indawgyi Lake is a localized pull-apart basin where a fault segment of the Sagaing fault steps to the adjacent fault segment on the right (Fig. 5). The east and west sides of the basin are bounded with strike-slip faults and north and south sides are bounded with irregular margins of normal faults. All these factors formed the background conditions of the geological structures which can generate the earthquakes.

4.5. The Pyinmana earthquake of 10 August 1931 was a violent earthquake which shook Pyinmana, 91 miles north of Phyu. It was felt as far north as Mandalay and was reported from Thanatpin in the Pegu district. The distance north and south over which it was felt was thus at least 350 miles. The occurrence of a converging bend for a large strike-slip fault like the Sagaing fault at Pyinmana area is the particular place where the stress

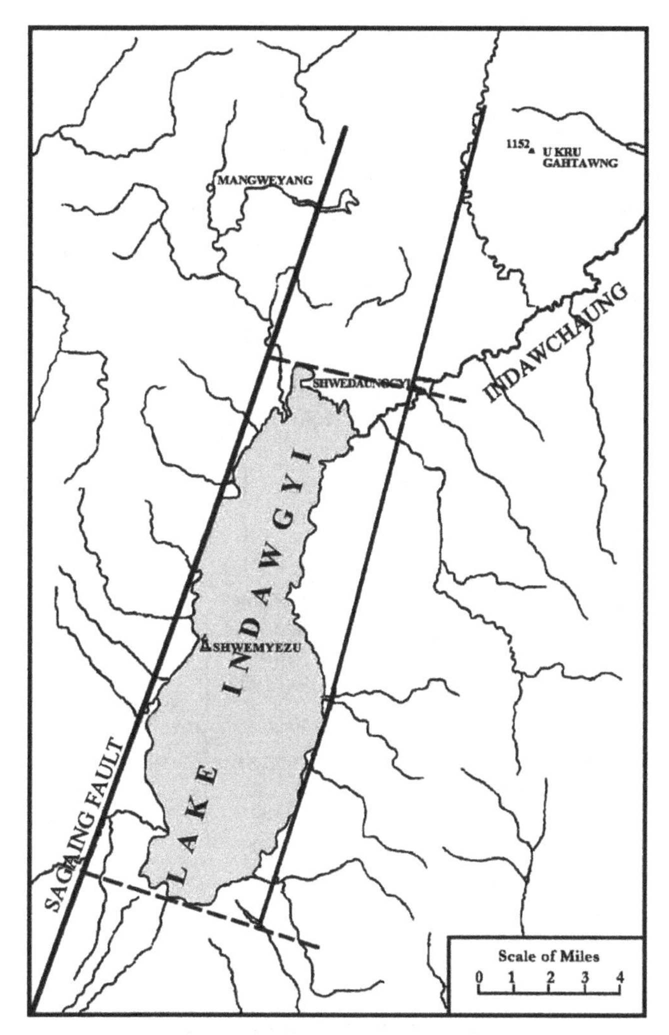

Fig. 5. Map of Indawgyi Lake (a localized pull-apart basin in the Sagaing fault zone, bounded with strike-slip faults and normal faults).[3]

accumulates and big earthquakes can occur. Such bends on the major strike-slip faults invite high concentration of strain.[12] It is a releasing bend where fault-bounded graben is formed by extensional deformation.

4.6. Phyu earthquake occurred on 23 September 1931. The shock brought about the collapse of the historic Shwe San Daw Pagoda, 165 feet high, at Taungoo.

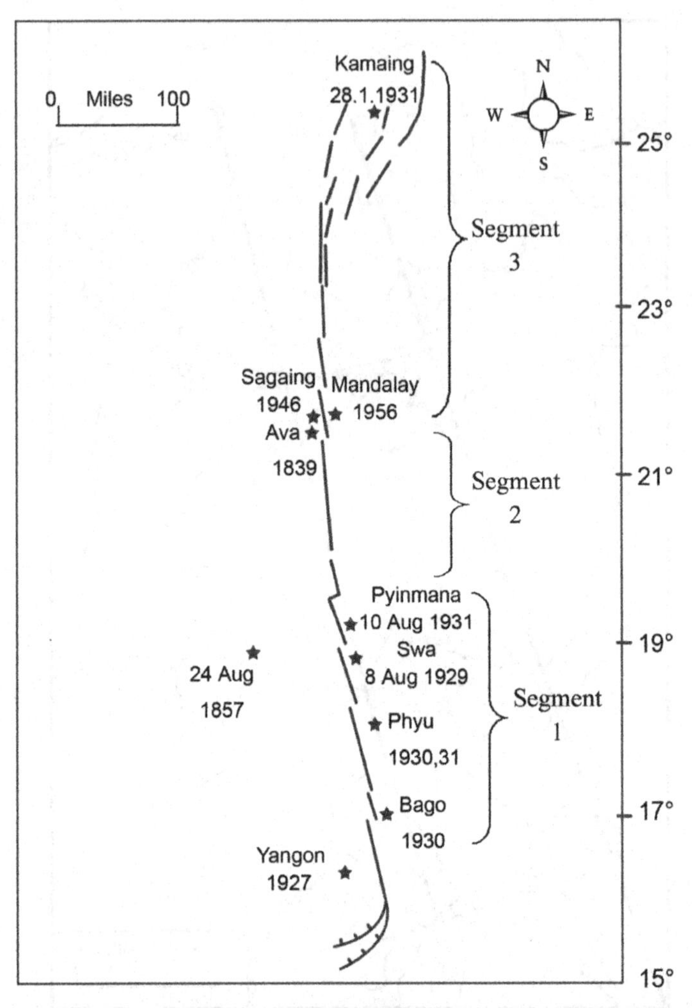

Fig. 6. The 1929–1931 sequence with six earthquakes released energy from two fault segments about 400 km length in the Segment (1) and 500 km length in the Segment (3) along the Sagaing fault.

Analysis of earthquake damages worldwide suggests that the severity of shaking depends on several local site-specific factors besides the distance and magnitude of an earthquake. Local site characteristics can lead to amplification of seismic waves and to unusually high damage. Certain frequencies of ground shaking may generate disproportionately large motion because of wave resonance and/or focusing in basins. Role of topography influence the intensity of seismic response at ridge-crests and de-amplification at ridge toes.[19] Several zones of aligned short fault segments arranged in right-stepping en-echelon pattern are found within the Sagaing fault zone. Between these fault segments of the fault zone, many localized pull-apart basins occur at fault bends near the tips of the fault segment of NNW-trending dextral faults. These basins like Kabauk In for Bago Earthquake, fault-bounded graben for Pyinmana Earthquake and Indawgyi Lake for Kamaing Earthquake and many other basins characterize step-overs where the intervening region is thrown into tension (Fig. 6). Linear fault scarps near Swa and Phyu areas within the Sagaing fault zone are tectonic geomorphic features which can lead to amplification of seismic response at ridge crests.

Conclusion

Among these earthquakes that occurred along the Sagaing fault, the 1929–1931 sequence with six earthquakes released energy from two fault segments about 400 km length in the Segment (1) and 500 km length in the Segment (3), along the same fault. Local site-characteristics and co-seismic deformation styles of these historical earthquakes showed that these events were dominated by normal faulting and strike-slip faulting mechanism. Because all the earthquakes for the period between 1929 and 1931 and their associated surface deformation provide two tectonic regimes shown as localized extensional basins and localized pressure ridges. Seismic activity appears to be related to NNW-oriented transtensional stress regime that is associated with ENE-oriented compression. Based on the interpretation of historical data, the earthquake amplification can only be the result of the co-seismic slip which usually nucleates near the tip of one fault segment and transfers to another segment for the strike-slip system, or the result of amplification of seismic waves at ridge crests. It is not only damaging buildings and pagodas, but also brings about permanent changes in the land and the fluvial system.

References

1. K. Abe, *Earth Planet. Inter.* **27** (1981) 72–92.
2. H. H. Aung, *Adv. Geosci.* **26** (2010).
3. F. Bender, *Geology of Burma* (Gebruder Borntraeger, BerlinStuttgart, 1983), 293 pp.
4. C. J. Brown and P. Leicester, *Mem. Geol. Surv. India* **62** (1993) 1–140.

5. C. J. Brown, P. Leicester and H. L. Chhibber, *Rec. Geol. Surv. India* **LXV** (1931) 227–284.
6. J. N. Brune, *J. Geophys. Res.* **73** (1968) 777–784.
7. H. L. Chhibber, *The Geology of Burma* (Macmillan and Co., London, 1934), 538 pp.
8. N. Christic-Blick and K. T. Biddle, *Soc. Econ. Pal. Mineral. Spec. Pub.* **37** (1985) 1–35.
9. J. C. Crowell, *Principles & Application of Terrane Analysis* (1983) 51–61.
10. J. R. Curray, *J. Asian Earth Sci.* **25** (2005) 187–232.
11. J. R. Curray, D. G. Moore, L. A. Lawver, F. J. Emmel, R. W. Raitt, M. Henry and R. Kieckhefer, *Am. Assoc. Petrol. Geol. Mem.* **29** (1979) 189–198.
12. G. H. Davies, *Structural Geology of Rocks and Regions* (John Wiley and Sons, Inc. Printed in United States of America, 1984).
13. S. Dasgupta, *Mem. Geol. Soc. India* **23** (1992) 319–334.
14. B. P. Dey, *J. Sci. Tech.* **1** (1968) 431–443.
15. T. J. Fitch, *J. Geophys. Res.* **77** (1972) 4432–4462.
16. B. Gutenberg and C. F. Richter, *Seismicity of the Earth and Associated Phenomena* (Princeton University Press, Princeton, N.J., 1954), 273 pp.
17. T. P. Harding, *Am. Assoc. Petrol. Geol. Bull.* **69** (1985) 582–600.
18. H. Kanamori, *J. Geophys. Res.* **82** (1977) 2981–2987.
19. P. Kearey, *Global Tectonics* (John Wiley and Sons, The Atrium, Southern Gate, Chichester, 2009).
20. P. K. Khan and P. P. Chakraborty, *Earth Planet. Sci. Lett.* **229** (2005) 259–271.
21. A. Y. Le Dain, P. Tapponnier and P. Molnar, *J. Geophys. Res.* **89** (1984) 453–472.
22. M. Van der Meijde and M. Shafique, *ITC News* (2010) 2–4.
23. C. Nielsen, N. Chamot-Rooke and C. Rangin, *Mar. Geol.* **209** (2004) 303–327.
24. D. A. Pivnik, J. Nahm, R. S. Tucker, G. O. Smith, K. Nyein, N. Nyunt, and P. H. Maung, *Am. Assoc. Petrol. Geol. Bull.* **82** (1998) 1837–1856.
25. K. A. K. Raju, T. Ramprasad, P. S. Rao, B. R. Rao and J. Varghese, *Earth Planet. Sci. Lett.* **7024** (2004) 1–18.
26. C. Rangin, Deformation of Myanmar, Results of GIAC Projects, *GIAC Conf.* (Yangon, Myanmar, 1996–1999).
27. B. Richter and M. Fuller, *Tectonic Evolution for Southeast Asia Geological Society Special Publication* No. 106 (1996) 203–224.
28. S. P. Satyabala, *Geophys. Res. Lett.* **25** (1998) 3189–3192.
29. S. P. Satyabala, *Int. Geophy.* **81** (2002) 797–798.
30. T. Sato, Master's Thesis: Hiroshima University, Higashi-Hiroshima, Japan, 2003.
31. A. Socquet, C. Vigny, N. Chamot-Rooke, W. Simons, C. Rangin and B. Ambrosius, *J. Geophys. Res.* **111** (2006) B05406, doi 10.1029/2005JB-003877.
32. R. C. Stein, G. C. P. King and J. Lin, *Science* **258** (1992) 1328–1332.
33. S. Stein and M. Wysession, *Introduction to Seismology, Earthquake, and Earth structure* (Blackwell Publishing Ltd. Victoria 3050, Australia, 2003).

34. H. Tsutsumi and T. Sato, *Bull. Seismol. Soc. Am.* **99**, 4 (2009) 2155–2168.
35. P. Tapponnier, G. Peltzer and R. Armijo, *Geological Special Publication* No. **19** (1986) 115–157.
36. C. Vigny, A. Socquet, C. Rangin, N. Chamot-Rooke, M. Pubellier, M.-N. Bouin, G. Bertrand and M. Becker, *J. Geophys. Res.* **108** (2003) ETG 6-1-10.
37. P. Wessel and W. H. F. Smith, *EOS* **79** (1998) 579.
38. W. Swe, *Contrib. Burmese Geol.* **1** (1981) 63–72.

Advances in Geosciences
Vol. 31: Solid Earth Science (2011)
Eds. Ching-Hua Lo *et al.*

THE IMPACTS OF COASTAL SUBSIDENCE AND SEA LEVEL RISE IN COASTAL CITY OF SEMARANG (INDONESIA)

H. Z. ABIDIN[*,**], H. ANDREAS[*], I. GUMILAR[*], Y. FUKUDA[†],
S. L. NURMAULIA[*,‡], E. RIAWAN[§], D. MURDOHARDONO[¶]
and SUPRIYADI[||]

[*] *Geodesy Research Group,*
Faculty of Earth Science and Technology,
Institute of Technology Bandung,
Jl. Ganesha 10, Bandung, Indonesia

[†] *Graduate School of Science,*
Kyoto University, Japan

[‡] *Institute of Physical Geodesy,*
Darmstadt University of Technology,
Petersenstrasse 13, D-64287, Darmstadt, Germany

[§] *Centre for Coastal and Marine Development,*
Institute of Technology Bandung,
Jl. Ganesha 10, Bandung, Indonesia

[¶] *Centre for Groundwater Resources*
and Environmental Geology,
Geological Agency of Indonesia, Bandung, Indonesia

[||] *Dept. of Physics, Semarang State University,*
Semarang, Indonesia
[**] *hzabidin@gd.itb.ac.id*

Semarang is the capital of Central Java province, located in the northern coast of Java island, which has a population of about 1.5 million. In general, the Leveling surveys, GPS surveys, Microgravity surveys, and InSAR technique, conducted from 1999 to 2011, determined that land subsidence in Semarang exhibits spatial and temporal variations, with the rates of about 1 to 19 cm/year. The largest subsidence occurred at several areas along the coast. This subsidence is mainly due to excessive groundwater extraction, natural consolidation of alluvial, and load of constructions. During the high tide periods, these subsiding areas used to experience flooding. This paper discusses the correlation between coastal subsidence and flooding phenomena along these coastal areas of Semarang, along with sea level rise phenomena in the Java sea. In this case, the sea level rise rates are obtained from altimetry observation.

1. Introduction

Semarang is the capital of Central Java province, located in the northern coast of Java island, Indonesia (see Fig. 1). It is centered at the coordinates of about $-6°58'$ (latitude) and $+110°25'$ (longitude), and covers an area of about 37,366.8 hectares or 373.7 km^2, with a population of about 1.43 million people in 2006.[11]

Topographically, Semarang consists of two major landscapes, namely the lowland and coastal area in the north and the hilly area in the south. The northern parts, where the city center, harbour, airport and railway stations are located, are relatively flat with topographical slopes ranging between 0° and 2°, and altitude between 0 and 3.5 m; while the southern parts have slopes up to 45° and altitude up to about 350 m above sea level. The northern parts have a relatively higher population density and also more industrial and business areas compared to the southern parts. The land use in the southern part consists of residential, office, retail, public use and open space areas. Two rivers run through the city, one on the east side and another on the west side which divides the city into three parts.

Geologically, Semarang has three main lithologies, namely, volcanic rock, sedimentary rock, and alluvial deposits. According to Ref. 13, the basement of Semarang consists of tertiary claystone of the Kalibiuk

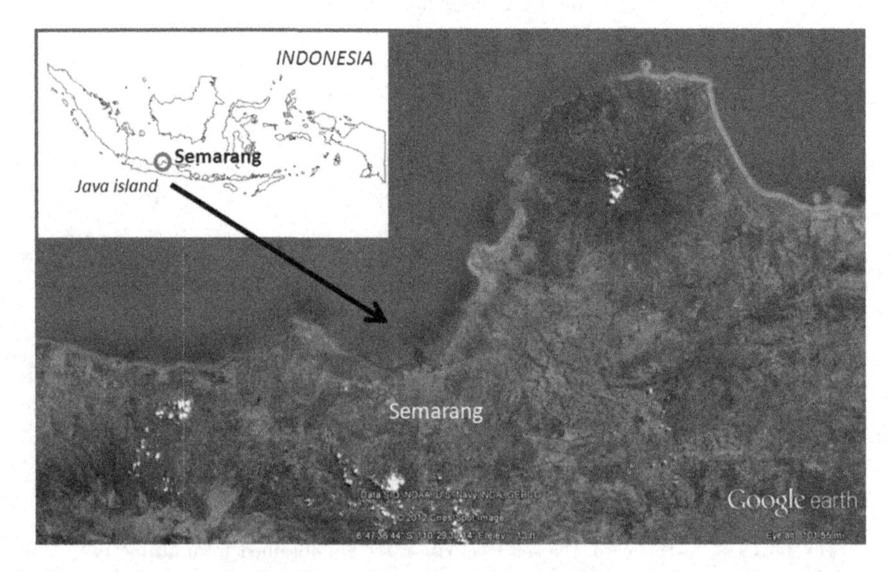

Fig. 1. Geographical location of Semarang; background image is from Google Earth.

formation. Overlying this formation is the Notopuro formation which consists of quaternary volcanic material. The two formations crop out in the southern part of the Semarang area. The northern part of the Semarang area is covered by Kali Garang deltaic alluvium up to a depth of 80 to 100 m in the coastal area. Aquifers are found at depths ranging from 30 to 80 m in this alluvium.

The northern part of Semarang is composed by very young alluvium with high compressibility. Several researchers[7,17] reported that the shoreline of Semarang progresses relatively quick toward the sea, namely about 2 km in 2.5 centuries or about 8 m/year. Therefore, it can be expected that natural consolidation process still occurs, causing land subsidence in the northern part of Semarang. Increases in the population and urban development in the area, has accelerated land subsidence through excessive groundwater extraction, and load of building and construction.

Land subsidence is not a new phenomenon for Semarang, which has experienced it for more than 100 years. The characteristics of land subsidence in Semarang have been studied using several geodetic methods, such as Leveling,[6,15] GPS surveys,[1] Microgravity,[4,12] and InSAR.[2,10] The impact of land subsidence in Semarang can be seen in several forms, such as the wider expansion of (coastal) flooding areas, cracking of buildings and infrastructure, and increased inland sea water intrusion. Land subsidence also badly influences the quality of living environment and life (e.g. health and sanitation condition) in the affected areas.

In the case of Semarang, comprehensive information on the characteristics of land subsidence is applicable to several important planning and mitigation efforts, such as effective control of coastal flood and seawater intrusion, spatial-based groundwater extraction regulation, environmental conservation, design and construction of infrastructure, and spatial development planning. Considering the importance of land subsidence information for supporting development activities in the Semarang area, monitoring and studying the characteristics of this subsidence phenomenon becomes more valuable.

The estimation using various observation methods reveal that subsidence rates in the coastal region of Semarang is relatively high, i.e. about 1 to 19 cm/year. The combined effects of this coastal subsidence and sea level rise will introduce other collateral hazards, namely the coastal (tidal) flooding phenomena. Several areas along the coast of Semarang already have experienced tidal flooding during high tide periods.

2. Land Subsidence Measurements in Semarang

Land subsidence in Semarang has been studied and estimated using several geodetic methods, namely, leveling survey, InSAR, microgravity survey, and GPS survey. Based on the leveling surveys conducted by the Center of Environmental Geology from 1999 to 2003, it was found that the relatively large subsidence were detected at around Semarang Harbor, Pondok Hasanuddin, Bandar Harjo and around Semarang Tawang Railway station, with the rates ranging from 1 to 17 cm/year.[9,16] Leveling results shows that the northern coastal areas of Semarang are subsiding with rates higher than 8 cm/year. These areas are generally composed of swamp deposits of soft clay.

The estimation based on the PS InSAR technique also revealed that areas close to the shoreline have subsidence rates of more than 8 cm/year.[5,10] This PS InSAR-based subsidence rates were derived from 28 ERS-2 and ENVISAT-ASAR radar scenes recorded between 27 November 2002 and 23 August 2006.

Land subsidence in Semarang has also been studied using the micro-gravity method since 2002 by the research group from the Department of Geophysics of ITB. Based on this method, it is found that during September 2002 to November 2005, a maximum subsidence of about 48 cm occurred in the northern region of Semarang.[14] It corresponds to a maximum rate of about 15 cm/year.

Land subsidence in Semarang was also estimated using GPS surveys.[1] With the GPS survey method, several monuments which are placed on the ground covering the Bandung Basin and its surroundings, are accurately positioned relative to a certain reference (stable) point, using the GPS survey technique. The precise coordinates of the monuments are periodically determined using repeated GPS surveys at specific time intervals. By studying the characteristics and rate of change in the height components of the coordinates from survey to survey, the land subsidence characteristics can be derived.

GPS surveys of land subsidence in Semarang have been conducted every year since 2008. The number of observed points is 52 points, in which their location and distribution of the points are shown in Fig. 2. Station SMG1 is the southernmost point in the network and considering its relatively stable location, it is used as the reference point for this subsidence study.

The GPS surveys exclusively used dual-frequency geodetic-type GPS receivers. The length of surveying sessions was between 9 and 11 h. The data were collected with a 30 s interval using an elevation mask of 15°.

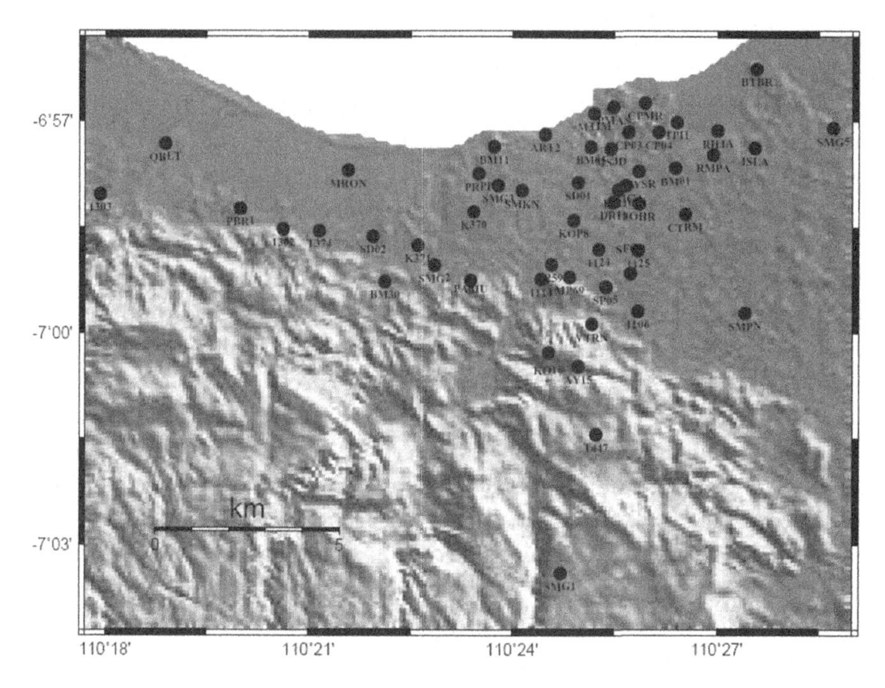

Fig. 2. GPS network for studying land subsidence in Semarang.

The surveys were mainly carried out by the staff and students from the Department of Geodesy and Geomatics Engineering of ITB (Institute of Technology Bandung).

The data were processed using the software Bernese 5.0.[3] Since we are mostly interested in the relative heights with respect to a stable point, the radial processing mode was used instead of a network adjustment mode. In this case, the relative ellipsoidal heights of all stations are determined relative to SMG1 station. For data processing, a precise ephemeris was used instead of the broadcast ephemeris. The effects of tropospheric and ionospheric biases are mainly reduced by the differencing process and the use of dual-frequency observations. The residual tropospheric bias parameters for individual stations are estimated to further reduce the tropospheric effects. The algorithms for the tropospheric parameter estimation can be found in Ref. 3. In processing baselines, most of the cycle ambiguities of the phase observations were successfully resolved. The standard deviations of GPS-derived relative ellipsoidal heights from all surveys were in general better than 1–2 mm. A few points have slightly larger standard deviations, due to the lack of observed data caused by the signal obstruction.

GPS derived subsidence rates as obtained from 2008 to 2011 are shown in Figs. 3–5, in the periods of (2008–2009), (2009–2010) and (2010–2011), respectively.

Results from GPS show that land subsidence in Semarang has spatial and temporal variations, ranging from 1 to about 19 cm/year, whereas the northern region of Semarang along the coastal area exhibits higher rates of subsidence compared to its southern region. Land subsidence results from leveling, InSAR, and microgravity methods, as shown in Figs. 6–8, also in general, reveal the same pattern of subsidence rates.

The relatively higher rate of land subsidence in northern part of Semarang is understandable, since this region is composed by very young alluvium with high compressibility. Therefore, it can be expected that natural consolidation process continued to occur until now, causing land subsidence on land above it. Increases in the population and urban development in the area, has accelerated land subsidence through excessive groundwater extraction, and load of building and construction.

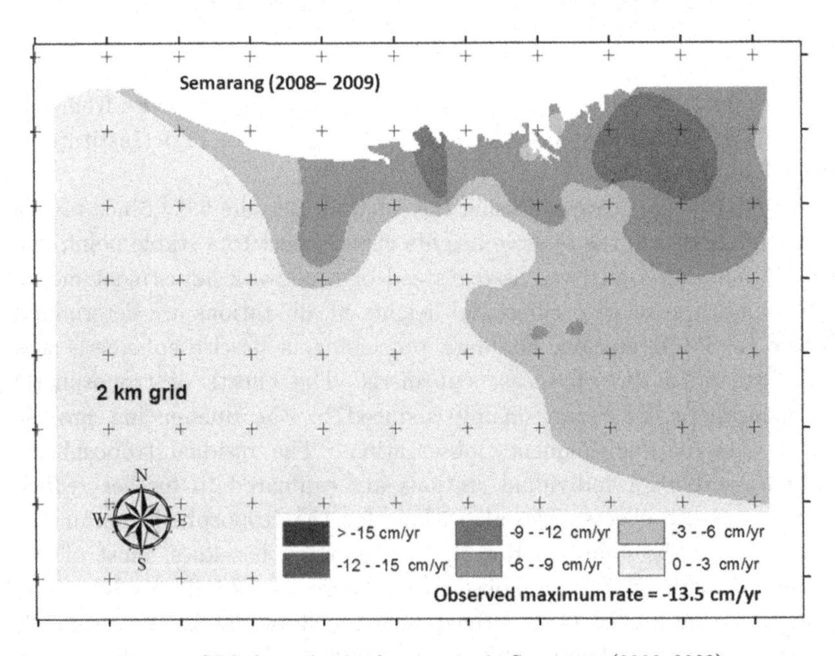

Fig. 3. GPS derived subsidence rates in Semarang (2008–2009).

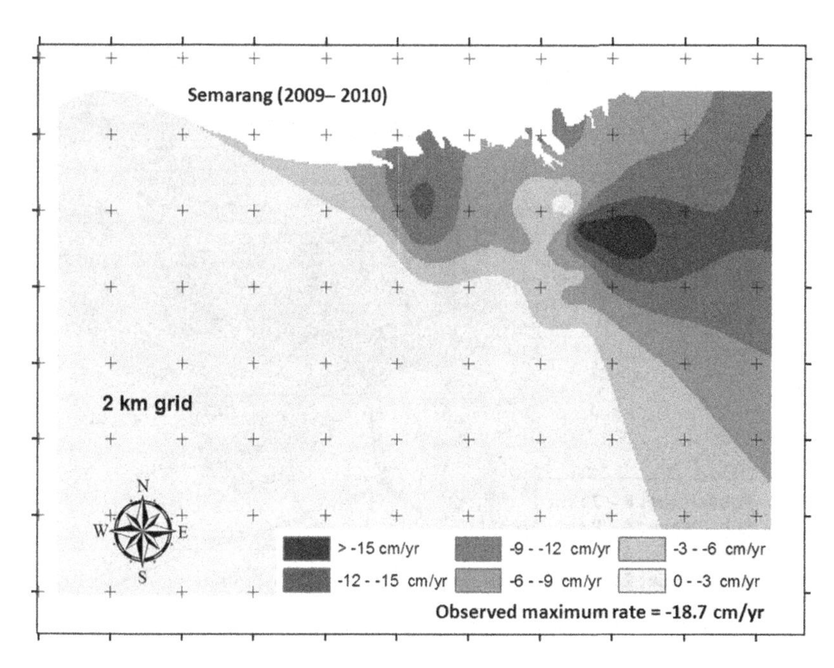

Fig. 4. GPS derived subsidence rates in Semarang (2009–2010).

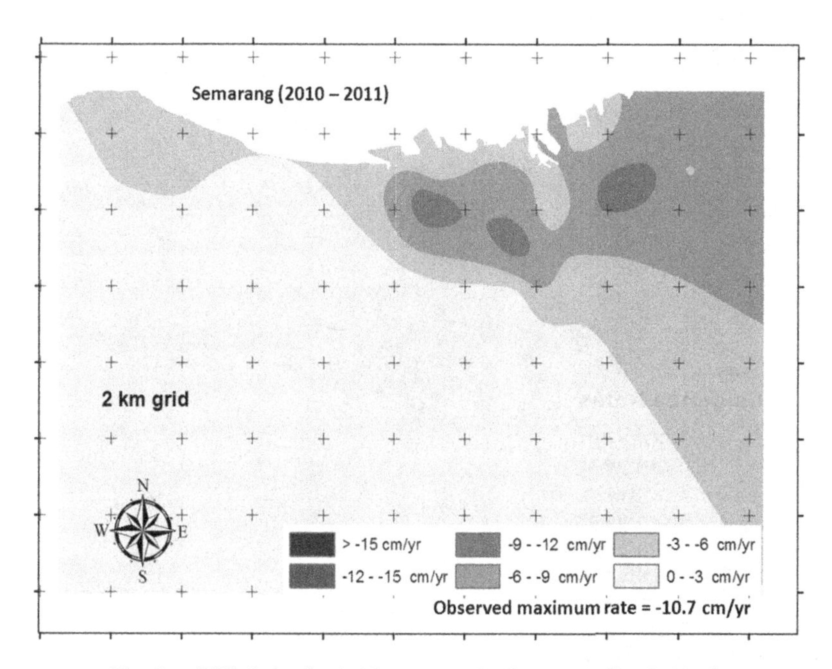

Fig. 5. GPS derived subsidence rates in Semarang (2010–2011).

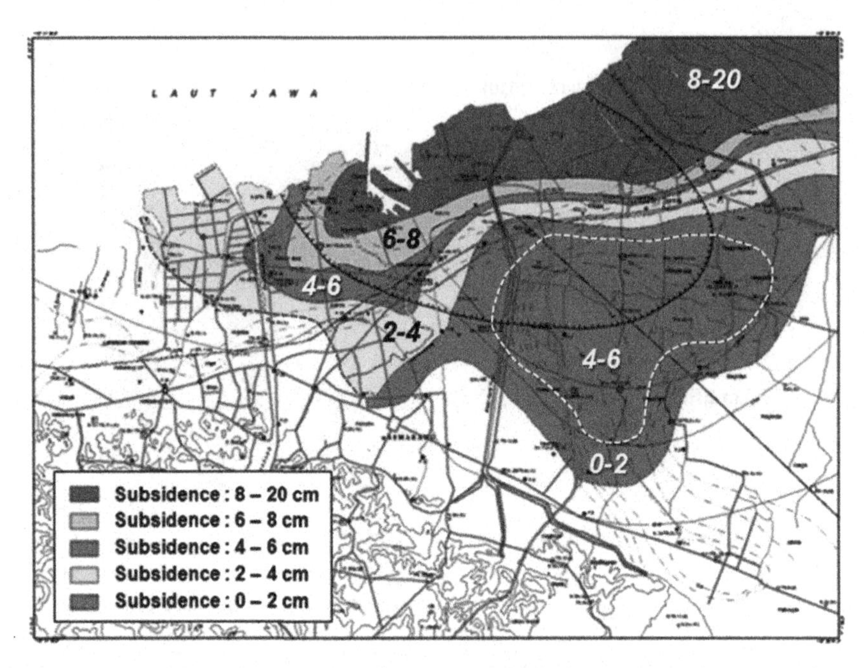

Fig. 6. Leveling derived subsidence in Semarang in the period of 2000 to 2001.[9]

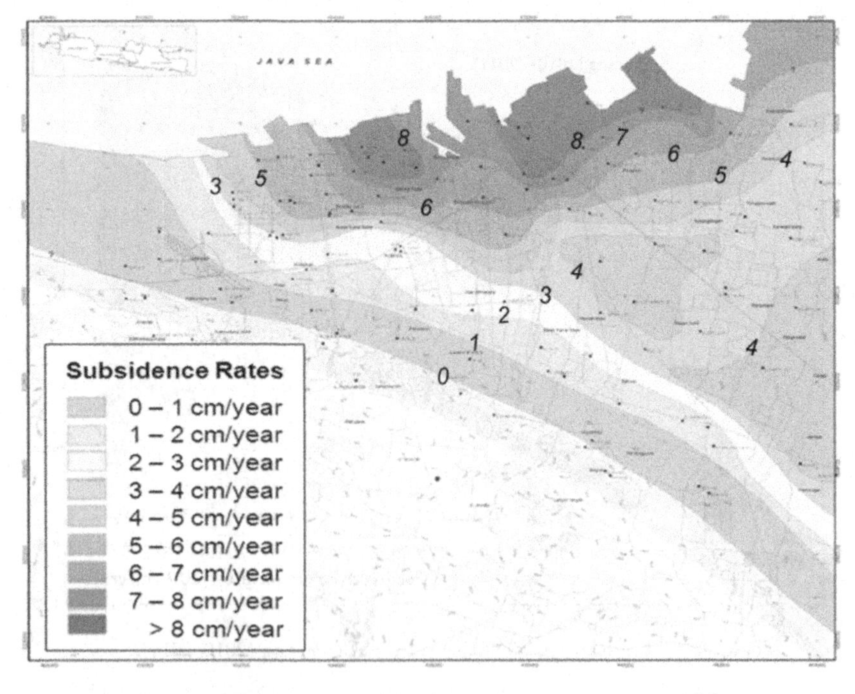

Fig. 7. PS InSAR derived subsidence rates in Semarang.[5,9]

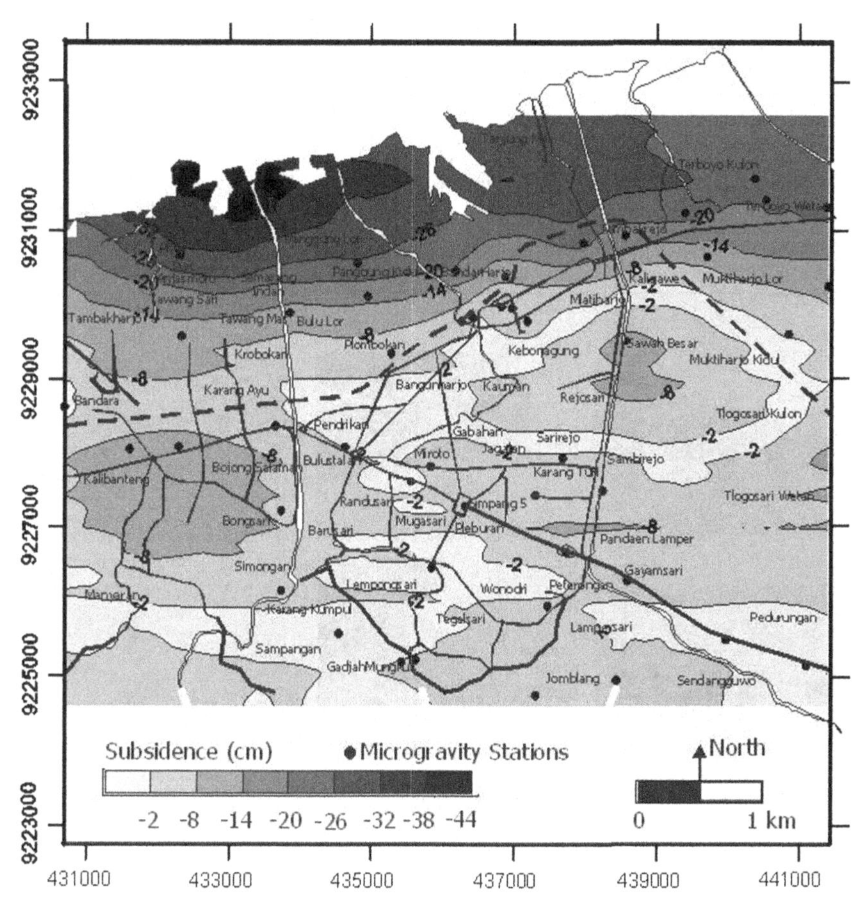

Fig. 8. Microgravity derived subsidence in Semarang from September 2002 to Nov. 2005.[14]

3. Impact of Land Subsidence in Semarang

Impact of land subsidence in the field can be seen in several forms, as summarized in Table 1. Land subsidence in Semarang has been widely reported and its impact can be seen already in daily life (see Fig. 9). It can be seen in the form of coastal flooding (it is called rob by the locals) where its coverage tends to enlarge manifolds. This frequent and severe rob not just deteriorates the function of building and infrastructures, it also badly influences the quality of living environment and life (e.g. health and sanitation condition) in the affected areas. Cracking of buildings and infrastructure, and increased inland sea water intrusion, are also other impacts of land subsidence.

Table 1. Several possible impact of land subsidence.

Impact of land subsidence		
Cracking of buildings and infrastructure	The wider expansion of inland and coastal flooding areas	Malfunction of drainage system
Increasing the maintenance costs for the affected buildings and infrastructure		Changes in river canal and drain flow systems
Lowering the quality of living environment and life (e.g. health and sanitation condition) in the affected areas		

Fig. 9. Examples of subsidence impact in Semarang.

The economic losses caused by land subsidence in Semarang are enormous; since many buildings and infrastructure in the industrial zone of Semarang are severely affected by land subsidence and its collateral coastal flooding disasters. Many houses, public utilities and a large population are also exposed to potential disaster. The corresponding maintenance cost is increasing by year. The provincial government and communities are required to frequently raise ground surface to protect roads and buildings. The living conditions, if the population affected by the land subsidence, are in general decreasing.

4. Land Subsidence and Sea Level Rise in Northern Coast of Semarang

The leveling, GPS, PS InSAR and microgravity derived results show that the coastal areas of Semarang are affected by subsidence phenomena with rates of about 1 up to 19 cm/year (see Table 2). During high tide,

Table 2. Land subsidence rates in Semarang as observed by different methods.[1,5,9,14]

No.	Method	Subsidence rates (cm/year)	Observation period
1	Leveling surveys	1–17	1999–2003
2	GPS surveys	1–19	2008–2011
3	PS InSAR	1 to more than 8	November 2002–August 2006
4	Microgravity surveys	1–15	September 2002–November 2005

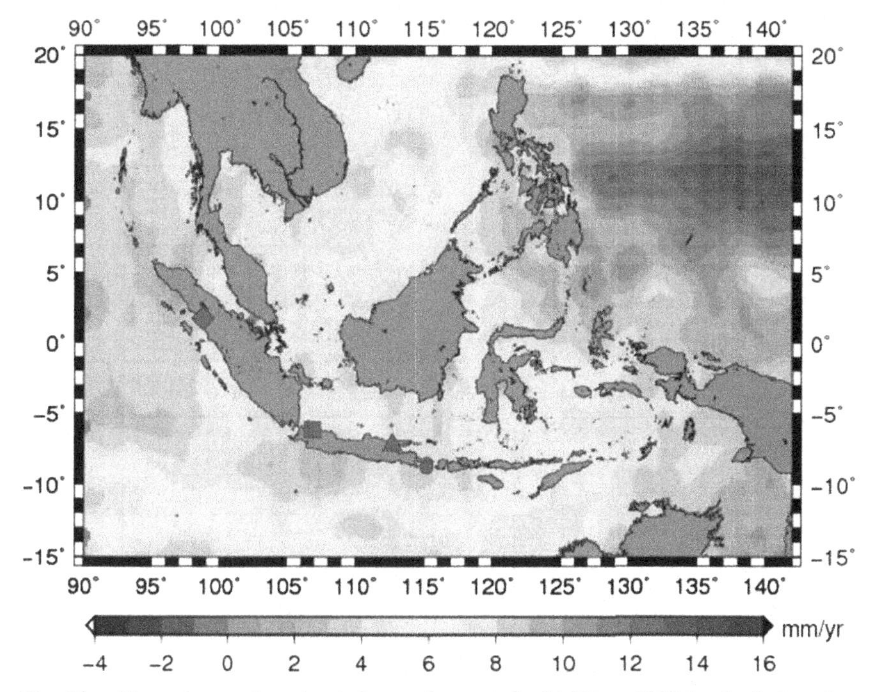

Fig. 10. Linear-term of sea level change for period of 1993 to 2009 in the Indonesian region from satellite altimetry.[18]

tidal flooding has been affecting most of the coastal areas. The extent and magnitude of subsidence related flooding will worsen with the likely continuation of sea level rise along the coastal area of Semarang, which is bordered by the Java sea.

Figure 10 shows the linear-term of sea level change for period of 1993 to 2009 in the Indonesian region as derived from satellite altimetry data.[18] Based on this figure, it can be seen that the sea level rise around Semarang coastal area is about 4–5 mm/year. This sea level rise rate is much less than the subsidence rates of the coastal land area of Semarang.

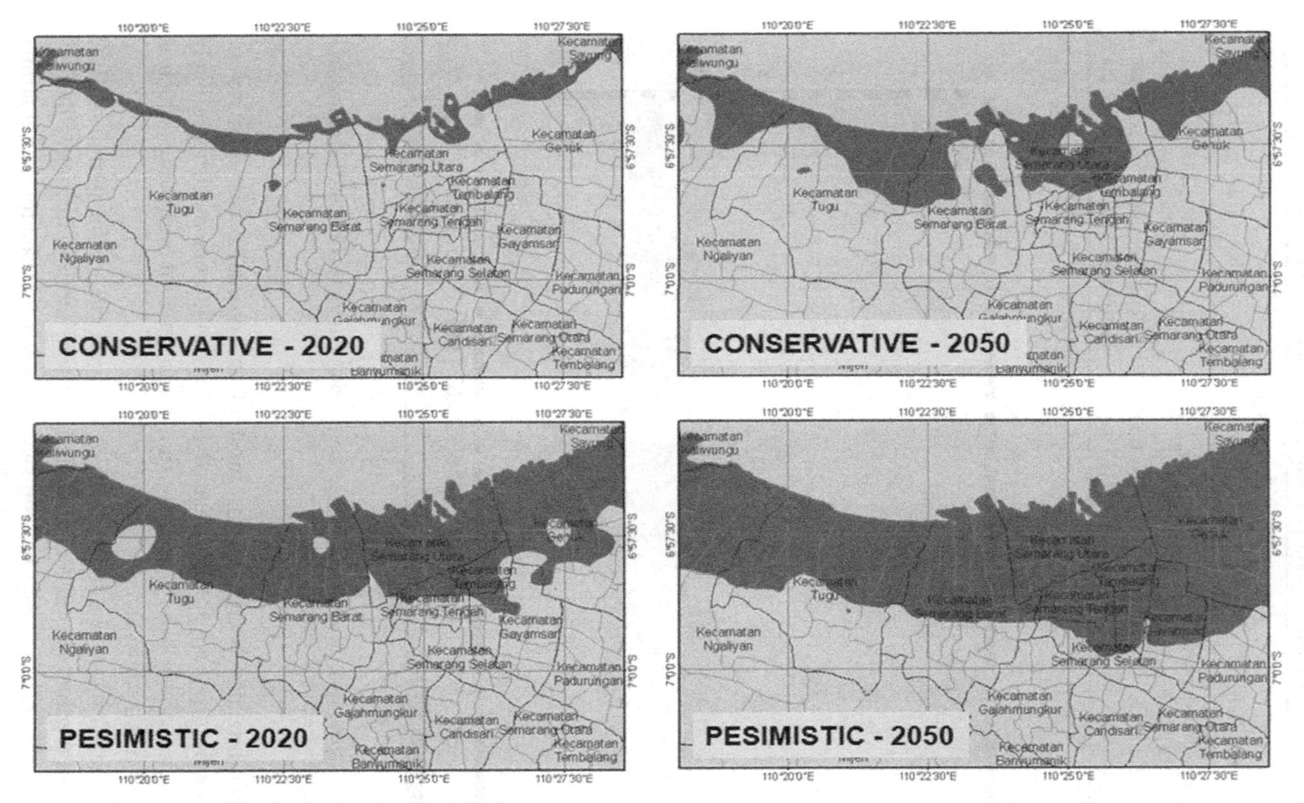

Fig. 11. Possible inundation areas (in black) in the coastal areas of Semarang; with the assumption of homogeneous subsidence rates along the coast.

Table 3. Scenarios for land subsidence and sea level rise rates in the coastal area of Semarang.

	Conservative scenario	Pesimistic scenario
Land subsidence rate	2.5 cm/year	10 cm/year
Sea level rise rate	0.2 cm/year	1 cm/year

The combined effects of land subsidence and sea level rise in the coastal areas of Semarang should be considered in vulnerability assessments of the areas to the tidal flooding phenomena. Table 2 shows two possible scenarios of future tidal flooding conditions, the first being a conservative estimate (most probable case) and the second being a pessimistic (worst case) scenario.

Considering the relatively flat nature (i.e. 0–2 m above MSL) of most coastal areas in Semarang, this combined effect of land subsidence and sea level rise will certainly have disastrous consequences for habitation, industry, and fresh groundwater supplies from the coastal aquifers. Figure 11 shows the possible inundated areas estimated using the scenarios given in Table 3.

It should be noted, however, that the land subsidence rate is not uniform over the entire coastal area of Semarang, as shown in Figs. 3–5. These figures show that some coastal areas are more susceptible to tidal flooding than the others. If the spatially different rates of subsidence as derived by GPS (see Fig. 3) are integrated with the scenarios of sea level rise rates given in Table 2, then the possible inundation maps as given in Fig. 12 are obtained.

5. Closing Remarks

Land subsidence in northern part of Semarang is believed to be caused by the combination of natural consolidation of young alluvium soil, groundwater extraction, and load of buildings and constructions. Due to the coastal land subsidence, part of the north coast area of Semarang city has been showing a growth of sea water inundation for nearly three decades.

Groundwater extraction in Semarang city is increasing sharply since the early 1990s, especially in the industrial area (see Fig. 13). Due to excessive groundwater extraction, the groundwater level in Semarang during the period of 1980 and 1996 is lowering with the rates of about 1.2 to 2.2 m/year.[18] This will then introduce land subsidence above it.

More data and further investigations, however, are required to understand the intricacies of the relationship between land subsidence and natural consolidation and groundwater extraction in Semarang area. Additional

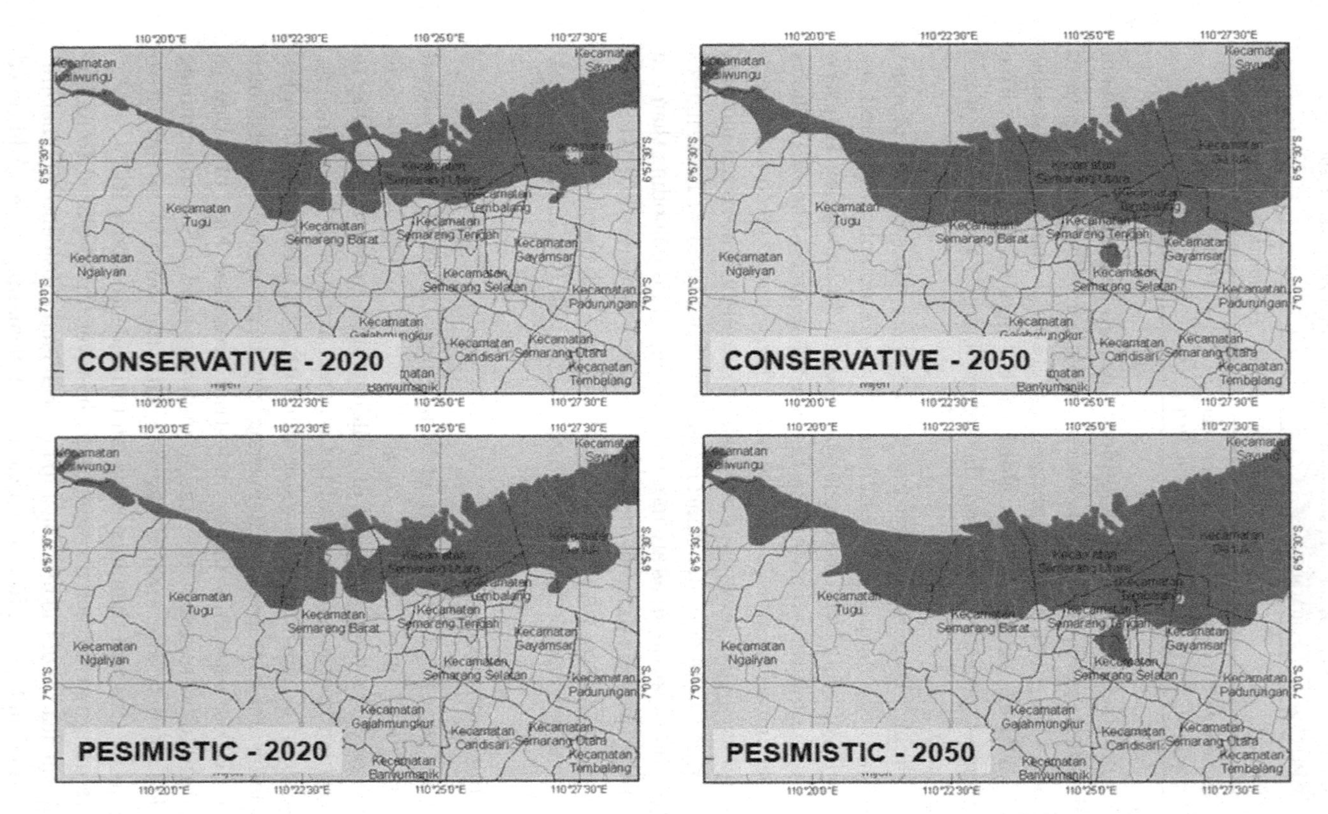

Fig. 12. Possible inundation areas (in black) in the coastal areas of Semarang; with the assumption of spatially different subsidence rates along the coast.

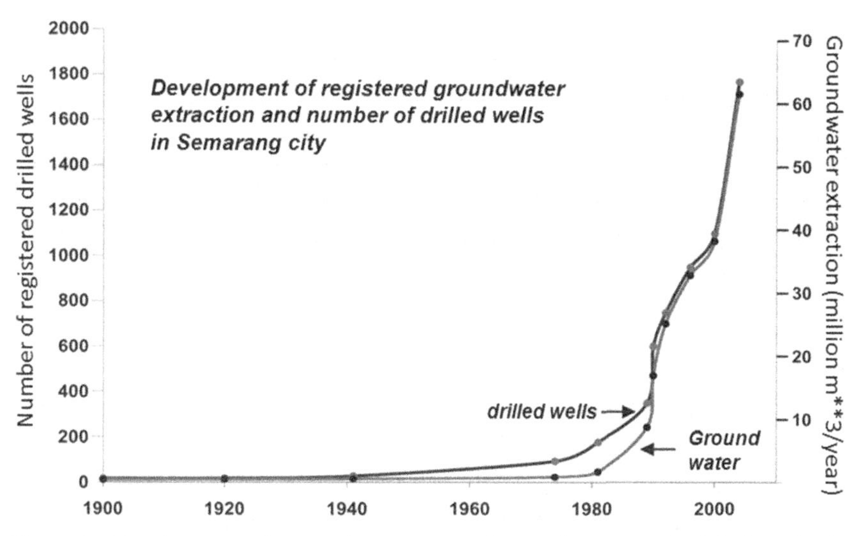

Fig. 13. Development of groundwater extraction and registered drilled wells in Semarang.[8]

Fig. 14. Coastal flooding in Semarang on mid April 2009; courtesy of Kompas photo, 2 July 2009.

causes of subsidence, e.g. load of buildings and construction, and tectonic movements, should also be investigated and taken into account.

Finally, it should be noted that in the coastal areas of Semarang, the combined effects of land subsidence and sea level rise will worsen the tidal flooding phenomena which has been already experienced by Semarang during the high tide periods, as shown by example in Fig. 14. Therefore, adaptation measures to reduce the impact of this hazard should be developed as soon as possible.

References

1. H. Z. Abidin, H. Andreas, I. Gumilar, T. P. Sidiq, M. Gamal, D. M. Supriyadi and Y. Fukuda, *Proceedings of the FIG Congress 2010*, Sydney, Australia (2010).
2. BGR, The site of Federal Institute for Geosciences and Natural Resources (BGR) of Germany. Available at: http://www.bgr.bund.de/ [accessed on 9 July 2009].
3. G. Beutler, H. Bock, R. Dach, P. Fridez, A. Gade, U. Hugentobler, A. Jaggi, M. Meindl, L. Mervant, L. Prange, S. Schaer, T. Springer, C. Urschl and P. Walser, *Bernese GPS software Version 5.0* (Astronomical Institute, University of Berne, 2007).
4. Y. Fukuda, T. Higashi, S. Miyazaki, T. Hasegawa, S. Yoshii, Y.Fukushima, J. Nishijima, M. Tanigushi, H. Z. Abidin and R. M. Delinom, *The AGU Fall Meeting*, San Fransisco, 15–19 December (2008).
5. F. Kuehn, D. Albiol, G. Cooksley, J. Duro, J. Granda, S. Haas, A. Hoffmann-Rothe and D. Murdohardono, *Environ. Earth Sci.* (2009), doi 10.1007/s12665-009-0227-x.
6. M. A. Marfai and L. King, *Environ. Geol.* **53** (2007) 651–659.
7. M. A. Marfai, H. Almohammad, S. Dey, B. Susanto and L. King, *Environ. Monit. Assess.* **142** (2008) 297–308.
8. Marsudi, PhD Dissertation, Institut Teknologi Bandung, (2001) 176 pp.
9. D. Murdohardono, T. M. H. L. Tobing and A. Sayekti, *The Int. Symp. and Workshop on Current Problems in Groundwater Management and Related Water Reosurces Issues*, Kuta, Bali, 3–8 December (2007).
10. D. Murdohardono, G. M. Sudradjat, A. D. Wirakusumah, F. Kühn, F. Mulyasari, The BGR-GAI-CCOP Workshop on Management of Georisks, 23–25 June, Yogyakarta (2009).
11. Semarang City, Oficial website of Semarang Munipality. Available at http://www.semarang.go.id/ [accessed on 1 July 2009].
12. M. Sarkowi, W. G. A. Kadir and D. Santoso, *Proceedings of the World Geothermal Congress 2005*, Antalya, Turkey 2005.
13. R. Sukhyar, *Atlas and Urban Geology* **14** (2003) 335–340.
14. Supriyadi, PhD Dissertation, Institute of Technology Bandung, September, (2008) 146 pp.

15. H. Sutanta, A. Rachman, Sumaryo and Diyono, *Proceedings of the Map Asia 2005 Conference*, Jakarta 2005.
16. TMHL Tobing and Dodid Murdohardono (2004). Site of Environmental Geology Agency. Available at: http://www.dgtl.esdm.go.id/ [accessed on 16 July 2004].
17. R. W. Van Bemmelen, *The Geology of Indonesia.* Vol. IA (Martinus Nijhof, The Netherlands, 1949).
18. S. L. Nurmaulia, FM. Luciana Fenoglio-Marc, M. Becker, *EGU General Assembly*, Vienna, Austria, 2010.

Advances in Geosciences
Vol. 31: Solid Earth Science (2011)
Eds. Ching-Hua Lo *et al.*

PHYSICO-CHEMICAL CHARACTERISTICS OF QUATERNARY SEDIMENTS OF TERNA RIVER SUB-BASIN EAST CENTRAL MAHARASHTRA, INDIA

Md. BABAR, R. V. CHUNCHEKAR and B. B. GHUTE
Department of Geology, Dnyanopasak College,
Parbhani-431 401, Maharashtra, India
md-babar@hotmail.com

Terna river valley is an important agricultural belt in east-central Maharashtra India. The study area, Terna river sub-basin a tributary river of Manjara river in Latur–Osmanabad districts, is considered for the present study. The physico-chemical characteristics of geomorphological units such as valley fill, pediments, pediplain, highly dissected plateau and older alluvial plain are discussed. The western part of the sub-basin is characterized by irregular hard rock terrain with flat-topped Deccan basaltic plateau surfaces, erosional and highly dissected hills. The eastern part of the sub-basin is gently sloping to flat terrain consisting of alluvial material of the Terna River and the E–W trending escarpment. The central and eastern part of the sub-basin is covered with thick black cotton soil and alluvial soil. The sediments in hilly areas of highly dissected plateau and pediment surfaces have proportions of sand greater than silt and clay. In pediplain and older alluvial soils, the proportion of clay and silt is greater than the sand and shows gradual decrease in the clay + silt to sand ratio. The pH ranges from 7.48 to 8.46, electric conductivity varies from 1.20 dS/m to 6.99 dS/m, organic carbon% from 0.02 to 1.15, $CaCO_3$% from 3 to 49.5 and potassium 10 to 12.4 ppm.

1. Introduction

The quaternary deposits in Marathwada area (India) are primarily fluvial and are confined to main river valleys as narrow belts. These deposits are generally discontinuous, unfossiliferous and lack suitable material for radiometric dating. Considering the lack of data on either the quaternary deposits or fluvial regime in Marathwada area, the Terna river basin has been selected for the present study. The study focuses on the physico-chemical characteristics of quaternary sediments of Terna river basin.

Quaternary fluvial sediments in the upland Deccan trap region have been studied by many researchers.[4–12] These sediments mostly occur in present day river valleys. Rajguru and Kale[9] divided quaternary sediments in five lithological units. They assigned a major portion of the exposed deposits to the late quaternary. Their chronology was based on several of ^{14}C dates obtained on freshwater molluscan shells and drift wood fragments. These streams were in an erosional mode during the early Holocene, due to a relatively strong southwest monsoon. Thus, it is clear that rivers particularly in western upland Maharashtra responded to global climatic changes during the late quaternary. Dole *et al.*[2] and Dole[3] argued against the climatic hypothesis of aggradational/erosional phases of the upland Rivers, particularly in the Pravara basin. Babar[1] determined the soil stratigraphy, morphostratigraphy and lithostratigraphy of quaternary sediments of Purna river sub-basin of Godavari River Basin, Maharashtra.

2. About the Area

Terna river valley is an important agricultural belt in east-central Maharashtra India. The study area, Terna river sub-basin a tributary river of Manjara river in Latur–Osmanabad districts, is bounded by latitude of 18° 25′ 48″ N and 18° 03′ 06″ N and longitude 75° 48′ 00″ E and 76° 57′ 0″ E (Fig. 1) and covers an area of 2,696.92 sq km. The area belongs to semiarid and subtropical climate zones characterized by hot summer and the normal annual rainfall is 802.40 mm.

3. Methodology

The various landforms and quaternary geomorphic and fluvial features like terraces are mapped in the field. The stratigraphic succession of the area is given in Table 1 and the lithostratigraphy of the quaternary sediments of the Terna basin is represented in Table 2. Numbers of samples collected from the different localities are analyzed for different physico-chemical parameters. The particle size distribution for percentages of sand, silt and clay is given in Table 3 and the physico-chemical parameters are presented in Table 4. The data collected represents different geomorphic surfaces including valley fill, pediments, pediplain, highly dissected plateau and denudational hills. The locations of sediments are shown in Fig. 1 and given Table 3.

Data generated in the field and lab are evaluated and synthesized to discuss the physico-chemical characteristics of quaternary sediments of the Terna river basin. The percentages of sand, silt and clay are analyzed

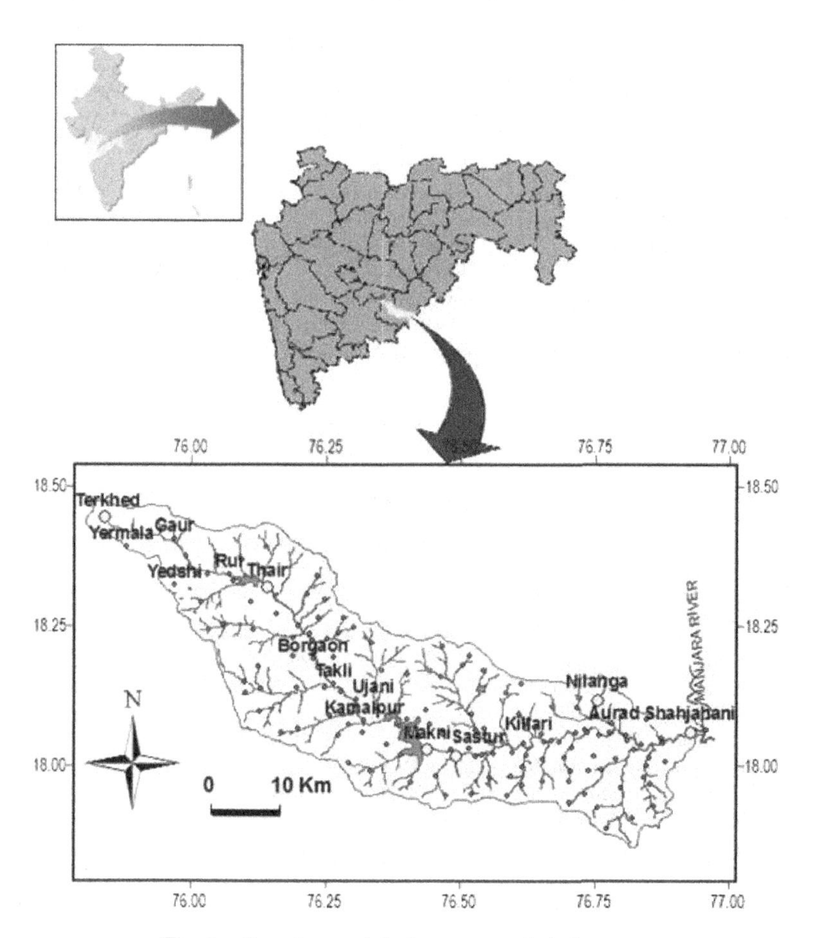

Fig. 1. Location and drainage map of study area.

Table 1. Stratigraphic succession of the study area.

Lithostratigraphy	Lithology	Soil characteristics and (type)	Morphostratigraphy
1. Dark grey silt formation	Uncalcified silty sand and gray silty clay	Gray, fine silty clay (Entisol)	T0 — Present Floodplain
2. Light grey silt formations	Light grey silty and brown to grey brown calcareous silty clay	Dark brown clayey (Vertisol)	T1 — Older Floodplain
3. Grey brown silt formation	Grey brown to brown calcareous clay	Dark brown clayey (Vertisol)	T2 — Pediplain

Table 2. Lithostratigraphy of quaternary formations of Terna river.

Era	Period	Formation	Lithology
Quaternary	Holocene	Dark grey silt formation in Present Floodplain (T0)	Uncalcified Silty Sand, Gray Silty Clay and Black Soil
	Late Upper Pleistocene	Light grey silt formations in Older Floodplain (T1)	Light Grey Silty and Brown to Grey Brown Calcareous Silty Clay
	Early Upper Pleistocene	Brown silt formation in Pediplain (T2)	Brown to Dark Brown Calcareous Clay
Tertiary and Mesozoic	Early Eocene to Late Cretaceous	Deccan Trap Basalt at base	Compact (Dense), Vesicular-Amygdaloidal Basalt Flows

Table 3. Particle size distribution of sediments.

			Soil particle distribution (in %)			
Sample no.	Locality	Geomorphic surfaces	Sand	Silt	Clay	Sediment type
TS1	Killari	Pediment	72.41	25.84	1.65	Sandy
TS2	Killari	Older alluvial plain	17.17	41.67	41.16	Clay Loamy
TS3	Sastur	Older alluvial plain	13.23	36.78	50.59	Silty Clay
TS4	Rajegaon	Pediplain	18.2	40.33	41.37	Silty Clay
TS5	Hattarga	Older alluvial plain	29.7	30.85	39.88	Silty Clay
TS6	Yelnur	Highly dissected plateau	47.36	31.88	20.63	Sandy silt
TS7	Yelnur	Pediment	66.5	26.79	6.78	Silty Sand
TS8	Ujni	Older alluvial plain	17.68	35.66	45.89	Silty Clay
TS9	Ujni	Present Floodplain	47.97	40.91	11.05	Silty Sand
TS10	Dhuta	Pediment	66.2	30.08	2.75	Silty Sand
TS11	Dhuta	Pediplain	28.35	43.55	25.98	Clay Loamy
TS12	Borgaon	Present Floodplain	47.09	36.55	17.35	Silty Sand
TS13	Borgaon	Pediment	90.2	9.1	0.69	Sandy
TS14	Ter	Older alluvial plain	31.3	30.64	37.62	Silty Clay
TS15	Yermala	Denudational hills	37.17	31.61	30.81	Silty Sand
TS16	Satephal	Pediment	53.49	26.66	20.23	Silty Sand
TS17	Satephal	Older alluvial plain	29.28	28.96	41.22	Silty Clay
TS18	Satephal	Present flood plain	88.86	9.54	1.37	Sandy
TS19	Rui	Highly dissected plateau	85.84	10.72	3.46	Sandy
TS20	Rui	Pediment	42.44	52.91	4.6	Sandy Silt
TS21	Sastur	Older alluvial plain	2.19	20.47	77.56	Clayey
TS22	Killari	Present flood plain	4.75	37.45	56.57	Silty Clay
TS23	Dhuta	Older alluvial plain	18.3	17.75	63.9	Silty Clay
TS24	Killari	Older alluvial plain	17.03	21.07	61.55	Silty Clay
TS25	Makhni	Older alluvial plain	22.49	22.36	55.31	Silty Clay

Table 4. Physico-chemical parameters of quaternary sediments.

Sample	Locality	pH	EC dS/m	Salinity ppm	Organic carbon %	Calciun carbonate %	Potassium K, ppm
TS1	Killari	8.03	2.20	169	0.39	14.5	10.2
TS2	Killari	7.70	2.15	96.5	1.15	11	10
TS3	Sastur	8.15	3.32	109	0.83	3	10.6
TS4	Rajegaon	7.80	3.35	115	0.31	10.5	10
TS5	Hattarga	7.95	2.43	121	0.41	6	10.1
TS6	Yelnur	7.71	2.48	130	0.53	19	10
TS7	Yelnur	8.12	2.08	104	0.43	33	11.2
TS8	Ujni	7.70	2.99	165	0.39	21	10
TS9	Ujni	7.95	2.42	118	0.39	35.5	10.6
TS10	Dhuta	8.24	1.53	81.2	0.41	37.5	11.2
TS11	Dhuta	8.02	2.01	109	0.51	32.5	12.4
TS12	Borgaon	7.93	1.94	87.2	0.45	28.3	10.1
TS13	Borgaon	8.44	1.20	63.8	0.04	49.5	10.4
TS14	Ter	8.46	5.39	274	0.70	10.5	10.1
TS15	Yermala	7.93	2.36	112	0.35	17	11.5
TS16	Satephal	7.90	4.12	210	0.25	22	10.6
TS17	Satephal	7.89	2.34	119	0.43	9.5	10
TS18	Satephal	8.12	1.70	86.9	0.41	20	10.1
TS19	Rui	8.19	1.79	86.7	0.35	29	10.8
TS20	Rui	8.35	2.75	134	0.02	49	11.3
TS21	Sastur	7.62	2..06	1.05	0.35	6	10.2
TS22	Killari	7.48	5.62	3.01	0.56	9	10.2
TS23	Dhuta	7.65	6.27	3.4	0.47	10.5	11.3
TS24	Killari	8.07	2.11	1.06	0.66	12.5	11.2
TS25	Makhni	7.6	6.99	372	0.39	13	10.3

EC = Electrical conductivity.

using the standard methods of grain size analysis. In addition, the pH and electrical conductivity (EC) are also determined. The percentage of organic carbon (OC) and the amount of $CaCO_3$ are estimated using standard methods and presented in Table 4.

4. Geology

Geologically, the entire study area is covered by Deccan basalt formations comprising gently sloping to nearly horizontal lava flows (Fig. 2). These flows have been considered to be a result of fissure type lava eruption during late Cretaceous to early Eocene period. The types of basalt flows occurring in the area are compact basalt flow (aa type) and vesicular-amygdaloidal basalt flows (pahoehoe type) as observed in the well sections.

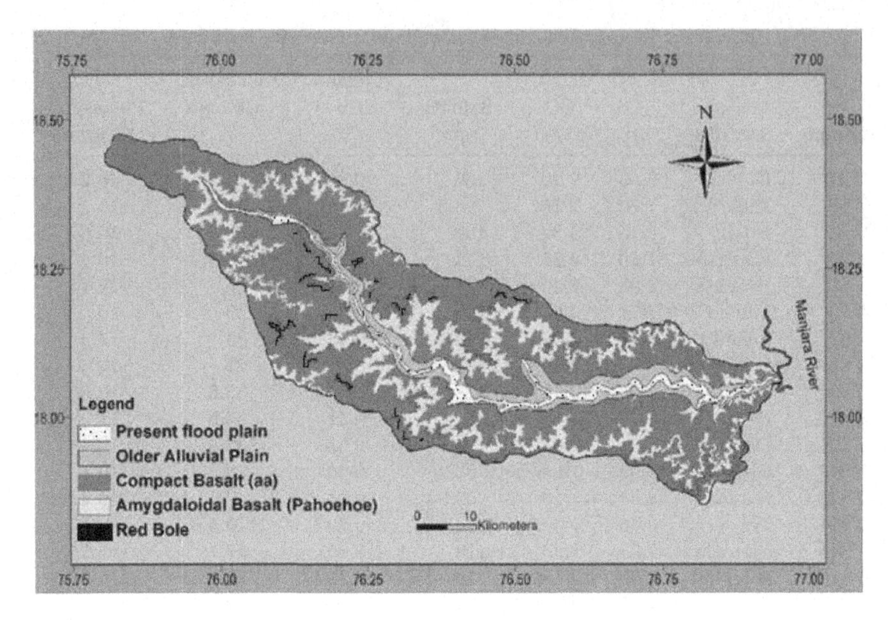

Fig. 2. Geological map of the Terna river basin.

The quaternary geological formations of the Terna river basin are described based on soil stratigraphy, morphostratigraphy and lithostratigraphy. Alluvial plain of the Terna river shows three terraces namely, T0, T1 and T2 in the increasing order. Besides the general stratigraphic succession of the area (Table 1), three lithostratigraphic formations (Table 2) are identified on the basis of order of superposition, nature of sediments, sedimentary structures and pedogenic characters. The deformational structures in the sediments include flexures, warps, buckle folds and vertical offset. Because of partitioning of mechanical behavior, the stiffer and more competent rocks are expected to show variation in shapes and wavelength of fold. From the style of deposition, it is clear that buckling of the sediment strata accompanied by flexural slip between the layers has formed the structures in sections at Ter, Killari, Sastur, Makni villages, etc.

5. Geomorphology

The physico-chemical characteristics of geomorphological units such as valley fill, pediments, pediplain, highly dissected plateau, and present and older flood surfaces are discussed. The western part of the sub-basin is characterized by irregular hard rock terrain with flat-topped Deccan basaltic plateau surfaces, erosional and highly dissected hills. The eastern part of the sub-basin is gently sloping to flat terrain consisting of alluvial material of Terna river and E–W trending escarpment. The central and eastern part of the sub-basin is covered with thick black cotton soil and alluvial soil.

The quaternary geological formations of Terna river basin are described based on soil stratigraphy, morphostratigraphy and lithostratigraphy. Alluvial plain of the Terna river shows three terraces namely, T0, T1 and T2 in the increasing order. Besides the general stratigraphic succession of the area, three lithostratigraphic formations have been identified on the basis of order of superposition, nature of sediments, sedimentary structures and pedogenic characters. The deformational structures in the sediments include flexures, warps, buckle folds and vertical offset in the sediments. Because of partitioning of mechanical behavior, the stiffer and more competent rocks are expected to show variation in shapes and wavelength of folds. From the style of deposition, it is clear that buckling of the sediment strata accompanied by flexural slip between the layers has formed the structures in sections at Ter, Killari, Sastur, Makni villages, etc.

6. Discussion

The physical characteristics studied included the particle size distribution with reference to the sand, silt and clay through the mechanical analysis of soil. The chemical parameters like pH, EC, OC, $CaCO_3$ were determined from the selected samples to understand the characteristics of the sediments in relation to processes and their environment. To establish the nature of sedimentation and processes involved, it was necessary to identify the different facies. This identification helped in understanding the geomorphic processes operating in time and space.

The data presented here was subjected to textural analysis. Textural analysis of the sediment is an important component particularly in the quaternary studies and agricultural sciences. Texture analysis estimates the different proportions of the sand, silt and clay in the sediments and puts them in a particular category, based on the proportion of sand, silt and clay. Particular classes of soils are favorable for plant growth, which was in turn related to water yielding and retention capacities, etc. The soils in the area are classified on the basis of color, texture, structure, consistency and nature of soil profile. Broadly, five types of sediments have been recognized on the basis of morphological characteristics such as soils at highly dissected plateau, pediment soils, pediplain soils, older alluvial and present flood plain soils.

In the hilly areas of highly dissected plateau and pediment surfaces, the proportions of sand is greater than silt and clay. In pediplain and older alluvial soils, the proportion of clay and silt is more than the sand, while the proportion of sand is decreasing. But in the present flood plains, the sand is greater and the silt and clay is decreasing. This study suggests that the present flood plain marks the vibrant geomorphic system and in other areas both the exogenic and endogenic processes are energetic and consequently the soil is converted in to clay.

The value of pH ranged from 7.48 to 8.46, electric conductivity varied from 1.20 dS/m to 6.99 dS/m, OC% from 0.02 to 1.15, $CaCO_3$% from 3 to 49.5 and potassium 10 to 12.4 ppm. The variation graph for EC versus pH (Fig. 3), OC% versus pH (Fig. 4) and $CaCO_3$% versus pH (Fig. 5) also revealed some interesting results.

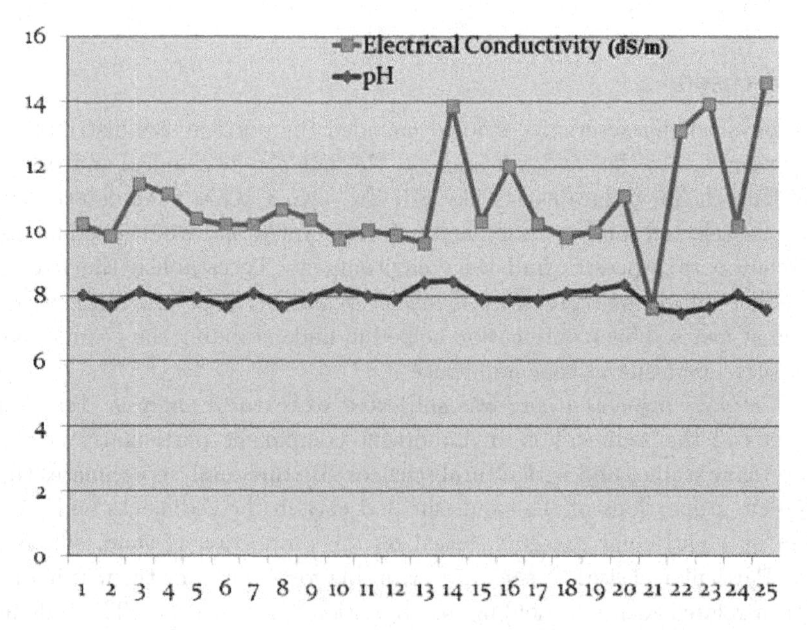

Fig. 3. Electrical conductivity and pH variations.

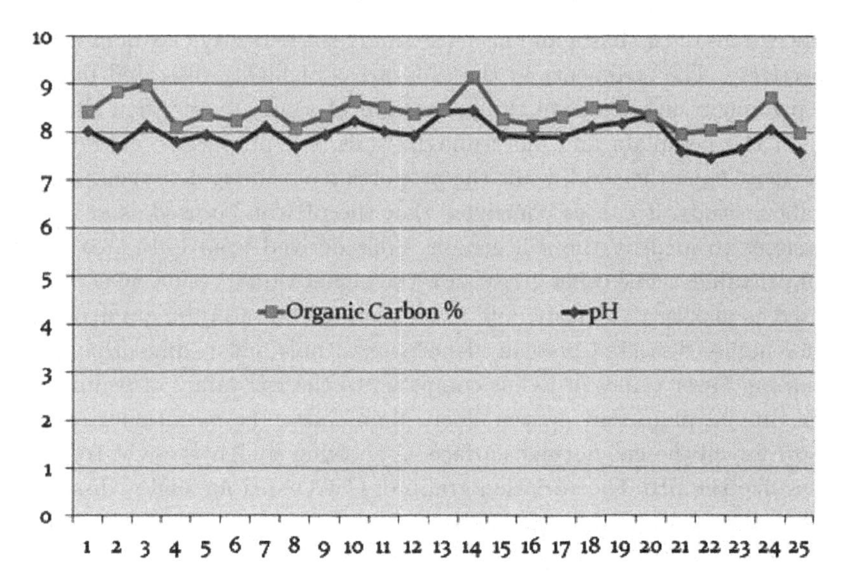

Fig. 4. Organic carbon and pH variation.

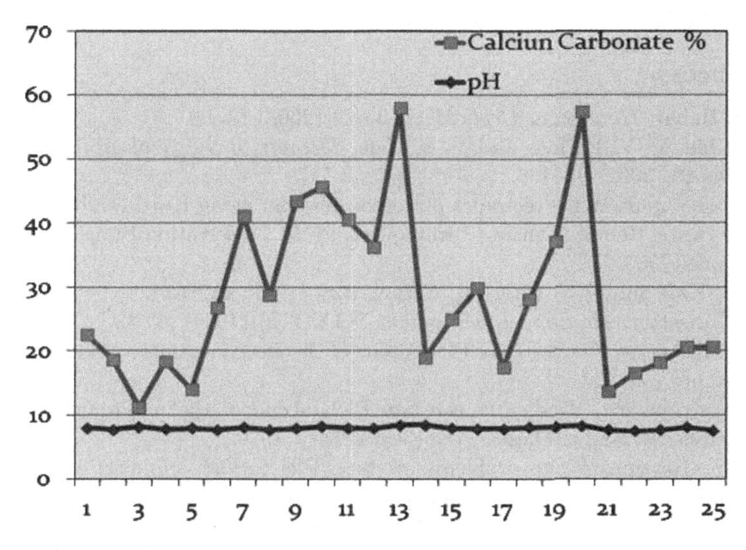

Fig. 5. CaCO3 and pH variation.

7. Conclusion

Geomorphic surfaces as in highly dissected plateau and hills have the problems of shallow soil cover, high relief, steep slope, rocky and rugged terrain difficult for agriculture. Pediplain and older alluvial plain are the fertile regions of the basin and have the sandy silt and silty clay dominance, respectively. The sediments in the hilly areas of highly dissected plateau and pediment surfaces have the proportion of sand greater than silt and clay. In the pediplain and older alluvial soils, the proportion of clay and silt is more than the sand, while the proportion of sand is decreasing. From the above study, it can be concluded that the pH can be used as an index parameter to identify the soil genesis. Soils derived from hilly area show lower pH values while other areas show the higher values. The $CaCO_3$% can be used as marker to identify different horizons. The variation graph of EC-pH for highly dissected plateau, denudational hills and pediment surfaces is showing lower values of EC as compared to the EC values of pediplains, older alluvial plain and present flood plain soils. The variation graph of OC-pH for all the geomorphic surfaces is showing similar straight trend of OC as like the pH. The variation graph of $CaCO_3$-pH for highly dissected plateau, denudational hills and pediment surfaces is showing higher trend of values of $CaCO_3$ as compared to the values of pediplains, older alluvial plain and present flood plain soils.

References

1. Md. Babar, *Gondwana Geol. Magz.* **23**(1) (2008) 81–90.
2. G. Dole, V. V. Peshwa and V. S. Kale, *Memoir of the Geological Society of India*, **49** (2002) 91–108.
3. G. Dole, Quaternary tectonics of deccan plateau along Kurduwadi lineament zone (using Remote Sensing Techniques), Ph.D. Dissertation Pune, University of Pune, 2005.
4. V. S. Kale and S. N. Rajguru, *Nature*, **325** (1987) 612–614.
5. R. Korisetter, *Man and Environment*, **XIX**(1–2) (1994) 29–42.
6. S. N. Rajaguru, On the Late Pleistocene of the Deccan Quaternaria, XI Roma, 1969, pp. 241–253.
7. S. N. Rajaguru, Studies in the late Pleistocene of the mula-mutha valley, Unpublished Ph.D. Thesis, Poona University, 1970.
8. S. N. Rajaguru, On problems of late Pleistocene, climatic changes in Western India, *Radiocarbon and Archeology* eds. D.P. Agrawal and A. Ghose (Tata Institute of Fundamental Research, Mumbai, 1973), pp. 80–87.
9. S. N. Rajaguru and V. S. Kale, *J. Geol. Soc. India*, **26** (1985) 16–27.
10. S. N. Rajaguru, V. S. Kale and G. L. Badam, *Current Sci.* **64**(11 and 12) (1993) 817–822.
11. S. N. Rajguru and R. Korisetter, *Indian Journal of Earth Sciences*, **14**(3–4) (1987) 817–822.
12. S. V. Umarjikar, Quaternary geology of the upper krishna basin, Ph.D. Dissertation, University of Poona, Pune, 1984.